Praise for *Context: Furt*
Productivity, Creativity, Pa
21st Ce

"As we live in the future going fas
that Cory Doctorow has given thought to the modern joys and dangers making our collective head spin. We all need to make time to have the conversations Cory starts in this book."
—Penn Jillette, co-star of Penn & Teller

"I can't say this about many authors, but I can say it about Cory, and without hesitation: Anyone who considers themselves smart, strategic, or even informed about where our digital economy is going (and I hope that's you) *must* read him. And this book is a great place to start."
—Seth Godin, author of *Linchpin*

"Reading *Context*, I felt like there should be a sticker on the cover, much like the one on Cracker Jacks, which promises us 'a prize in every box' or perhaps the old slogan for Lay's Potato Chips, 'bet you can't eat just one!' These bite-sized clusters of observations are munchable and easy to digest, but inside, they carry thoughts that can wake you up in the middle of night. The topics here range across intellectual property, science fiction, technological innovation, media policy, and electronic publishing, but he is often at his best when he pulls things down to the human level, describing the pleasures of being a parent in the digital age, or that guy he knew long ago who wore his sweaters inside out."
—Henry Jenkins, author of *Convergence Culture: Where Old and New Media Collide*

"Cory Doctorow thinks about lots of things, and he writes about lots of things, and he does both in a way that sends

some folks right over the edge. It's not that Cory is being outrageous to be outrageous—it's that he realizes that the context of our lives is change. That's a message some people don't want to hear. Well, I want to hear it. I don't always agree with Cory 100%—who agrees with anyone else all the time?—but I never get tired of reading what he's thinking about next."
—John Scalzi, author of *Old Man's War* and *Fuzzy Nation*

"See the emerging world of electronic books, iPad apps, cloud computing, and more through the eyes of possibly the most productively opinionated commentator of our day. Any complacency you have about the digital everyday will not survive unscathed."
—Mizuko Ito, Professor and MacArthur Foundation Chair in Digital Media and Learning, University of California, Irvine

Praise for *Content: Selected Essays on Technology, Creativity, Copyright, and the Future of the Future*

"Doctorow here proves he's smart, funny, and good at accessibly boiling down issues he's passionate about...a pleasure to read, not to mention thought-provoking."
—*Booklist*

"...more than just insightful, brilliant, and to the point—it's also funny and fun to read."
—Electronic Frontier Foundation

"If you want to know what's happening at the sharp end of digital publication and new ideas about the relationships between authors and their readers—do yourself a favour and listen to what he has to say."
—Mantex Online

tachyon / san francisco

Cover and interior design by Elizabeth Story
Author photo © 2005 by Patrick H. Lauke aka Redux (*www.splintered.co.uk*)

Tachyon Publications
1459 18th Street #139
San Francisco, CA 94107
(415) 285-5615
www.tachyonpublications.com
tachyon@tachyonpublications.com

Series Editor: Jacob Weisman
Project Editor: Jill Roberts

ISBN 13: 978-1-61696-048-3
ISBN 10: 1-61696-048-5

Printed in the United States of America by Worzalla

First Edition: 2011

9 8 7 6 5 4 3 2 1

For EFF's founders:
Mitch Kapor, John Perry Barlow, and John Gilmore,
who understood before the rest of us

Foreword
Tim O'Reilly

Edwin Schlossberg once said "The skill of writing is to create a context in which other people can think." And oh, how we need that skill today!

In times of transition and upheaval, we are literally "off the map" of past experience that is our normal guide to what to expect and how to think about it. It's at times like these that we need context-setters to shape how we understand and think about the changes facing us.

It was clear from the first that Cory Doctorow is one of the great context-setters of our generation, helping us all to understand the implications of the technology being unleashed around us. We are fortunate that unlike many who practice this trade, who look backward at recent changes, or forward only a year or two, Cory uses the power of story to frame what is going on in larger terms.

From his first novel, *Down and Out in the Magic Kingdom*, to his latest, *For the Win*, Cory helps us make sense of the world that is unfolding. The ideas behind his stories are *tools to think with* about hard problems in futures few are even prescient enough to predict. What kind of economy might we build when physical goods are virtually free? Might we see labor unions in MMORPGs? How might young adults foil the surveillance society?

Like Cory, I live in the future, or what might appear to be the future to those who aren't yet aware of how the world has already moved on. I am surrounded by software developers, innovators, and entrepreneurs, each of whom is building elements of a new world. Yet even those who are at the cutting edge of technology need a context to think in. It's easy for them to get caught up in trivialities—in building the next generation of consumer applications, in creating shiny toys rather than services of enduring value.

And it's here that Cory's profound moral sense comes to the fore. He is passionate about the potential of technology to build a better world, and evangelical about our responsibility to make it so.

And if each of Cory's novels and stories is packed with insight about possible futures, his essays are, if possible, an even more pure dose. Here is your chance to see a humane and thoughtful mind coming to grips with life as it is now, and as it is becoming.

Cory's writing is didactic in the best sense. Each of his stories or essays teaches us something, often many things, about the world to come and what we need to know to survive and prosper in it. They teach entertainingly, but they do teach. Are you ready to learn?

Jack and the Interstalk: Why the Computer Is Not a Scary Monster

With a little common sense, parents have nothing to fear from letting young children share their screen time

"Daddy, I want something on your laptop!" These are almost invariably the first words out of my daughter Poesy's mouth when she gets up in the morning (generally at 5 a.m.). Being a lifelong early riser, I have the morning shift. Being a parent in the 21st century, I worry about my toddler's screen time—and struggle with the temptation to let the TV or laptop be my babysitter while I get through my morning email. Being a writer, I yearn to share stories with my two-year-old.

I can't claim to have found the answer to all this, but I think we're evolving something that's really working for us—a mix of technology, storytelling, play, and (admittedly) a little electronic babysitting that let's me get to at least *some* of my email before breakfast time.

Since Poe was tiny, she's climbed up on my lap and shared my laptop screen. We long ago ripped all her favorite DVDs (she went through a period at around 16 months when she delighted in putting the DVDs shiny-side-down on the floor, standing on them, and skating around, sanding down the surface to a perfectly unreadable fog of microscratches). Twenty-some movies, the whole run of *The Muppet Show*, some BBC nature programmes. They all fit on a 32GB SD card and my wife and I both keep a set on our laptops for emergencies, such as in-flight meltdowns or the occasional restaurant scene.

I use a free/open source video player called VLC, which plays practically every format ever invented. You can tell it to eliminate all its user interface, so that it's just a square of movable video, and the Gnome window-manager in Linux lets me set that window as "Always on top." I shrink it down to a postage stamp and slide it into the top right corner of my screen, and that's Poesy's bit of my laptop.

When she was littler, we'd do this for 10 or 20 minutes every morning while she went from awake to awake-enough-to-play. Now that she's more active, she usually requests something—often something from YouTube (we also download her favourite YouTube clips to our laptops, using deturl.com), or she'll start feeding me keywords to search on, like "doggy and bunny" and we'll have a look at what comes up. It's nice sharing a screen with her. She points at things in her video she likes and asks me about them (pausable video is great for this!), or I notice stuff I want to point out to her. At the same time, she also looks at my screen—browser windows, email attachments, etc.—and asks me about them, too.

But the fun comes when we incorporate all this into our storytelling play. It started with Jack and the Beanstalk. I told her the story one morning while we were on summer vacation. She loved the booming FEE FI FOE FUM! but she was puzzled by unfamiliar ideas like beanstalks, castles, harps, and golden eggs. So I pulled up some images of them (using Flickr image search). Later, I found two or three different animated versions of Jack's story on YouTube, including the absolutely smashing Max Fleischer 1933 version. These really interested Poesy (especially the differences between all the adaptations), so one evening we made a Lego beanstalk and had an amazing time running around the house, play-acting Jack and the Beanstalk with various stuffed animals and such as characters. We made a golden egg out of wadded up aluminium foil, and a harp out of a coat-hanger, tape, and string, and chased up and down the stairs bellowing giant-noises at one another.

Then we went back to YouTube and watched more harps, made sure to look at the geese the next Saturday at Hackney City Farm, and now every time we serve something small and bean-like with a meal at home, there's inevitably a grabbing up of two or three of them and tossing them out the window while shouting, "Magic beans! Magic beans! You were supposed to sell the cow for money!" Great fun.

Every parent I know worries about the instantaneously mesmerizing nature of screens for kids, especially little kids. I've heard experts advise that kids be kept away from screens until the age of three or four, or even later, but that's not very realistic—at least not in our house, where the two adults do a substantial amount of work, socialising, and play from home on laptops or consoles.

But the laptop play we've stumbled on feels *right*. It's not passive, mesmerised, isolated TV watching. Instead, it's a shared experience that involves lots of imagination, physically running around the house (screeching with laughter, no less!), and mixing up story-worlds, the real world, and play. There are still times when the TV goes on because I need 10 minutes to make the porridge and lay the table for breakfast, and I still stand in faint awe of the screen's capacity to hypnotise my toddler, but I wouldn't trade those howling, hilarious, raucous games that our network use inspires for anything.

Teen Sex

My first young adult novel, *Little Brother*, tells the story of a kid named Marcus Yallow who forms a guerilla army of young people dedicated to the reformation of the U.S. government by any means necessary. He and his friends use cryptography and other technology to subvert security measures, to distribute revolutionary literature, to liberate and publish secret governmental memos, and humiliate government officials. Every chapter includes some kind of how-to guide for accomplishing this kind of thing on your own, from tips on disabling radio-frequency ID tags to beating biometric identity system to defeating the censorware used by your school network to control what kind of things you can and can't see on the internet. The book is a long hymn to personal liberty, free speech, the people's right to question and even overthrow their government, even during wartime.

Marcus is 17, and the book is intended to be read by young teens or even precocious tweens (as well as adults). Naturally, I anticipated that some of the politics and technology in the story would upset my readers. And it's true, a few of the reviewers were critical of this stuff. But not many, not overly so.

What I didn't expect was that I would receive a torrent of correspondence and entreaties from teachers, students, parents, and librarians who were angry, worried, or upset that Marcus loses his virginity about two-thirds of the way through the book (secondarily, some of them were also offended by the fact that Marcus drinks a beer at one point, and a smaller minority wanted to know why and how Marcus could get away with talking back to his elders).

Now, the sex-scene in the book is anything but explicit. Marcus and his girlfriend are kissing alone in her room after a climactic scene in the novel, and she hands him a condom. The scene ends. The next scene opens with Marcus reflecting that it wasn't what he thought it would be, but it was still very good, and better in some ways than he'd expected. He and his girlfriend have been together for quite some time at this point, and there's every indication that they'll go on being together for some time yet. There is no anatomy, no grunts or squeals, no smells or tastes. This isn't there to titillate. It's there because it makes plot-sense and story-sense and character-sense for these two characters to do this deed at this time.

I've spent enough time explaining what this "plot-sense and story-sense and character-sense" means to enough people that I find myself creating a "Teen transgression in YA literature FAQ."

There's really only one question: "Why have your characters done something that is likely to upset their parents, and why don't you punish them for doing this?"

Now, the answer.

First, because teenagers have sex and drink beer, and most of the time the worst thing that results from this is a few days of social awkwardness and a hangover, respectively. When I was a teenager, I drank sometimes.

I had sex sometimes. I disobeyed authority figures sometimes.

Mostly, it was OK. Sometimes it was bad. Sometimes it was wonderful. Once or twice, it was terrible. And it was thus for everyone I knew. Teenagers take risks, even stupid risks, at times. But the chance on any given night that sneaking a beer will destroy your life is damned slim. Art isn't *exactly* like life, and science fiction asks the reader to accept the impossible, but unless your book is about a universe in which disapproving parents have cooked the physics so that every act of disobedience leads swiftly to destruction, it won't be very credible. The pathos that parents would like to see here become bathos: mawkish and trivial, heavy-handed, and preachy.

Second, because it is good art. Artists have included sex and sexual content in their general-audience material since cave-painting days. There's a reason the Vatican and the Louvre are full of nudes. Sex is part of what it means to be human, so art has sex in it.

Sex in YA stories usually comes naturally, as the literal climax of a coming-of-age story in which the adolescent characters have undertaken a series of leaps of faiths, doing consequential things (lying, telling the truth, being noble, subverting authority, etc.) for the first time, never knowing, *really* knowing, what the outcome will be. These figurative losses of virginity are one of the major themes of YA novels—and one of the major themes of adolescence—so it's artistically satisfying for the figurative to become literal in the course of the book. This is a common literary and artistic technique, and it's very effective.

I admit that I remain baffled by adults who object to the sex in this book. Not because it's prudish to object, but because the off-camera sex occurs in the middle of a story that features rioting, graphic torture, and detailed instructions for successful truancy.

As the parent of a young daughter, I feel strongly that every parent has the right and responsibility to decide how his or her kids are exposed to sex and sexually explicit material.

However, that right is limited by reality: the likelihood that a high-school student has made it to her 14th or 15th year without encountering the facts of life is pretty low. What's more, a kid who enters puberty without understanding the biological and emotional facts about her or his anatomy and what it's for is going to be (even more) confused.

Adolescents think about sex. All the time. Many of them have sex. Many of them experiment with sex. I don't believe that a fictional depiction of two young people who are in love and have sex is likely to impart any new knowledge to most teens—that is, the vast majority of teenagers are apt to be familiar with the existence of sexual liaisons between 17-year-olds.

So since the reader isn't apt to discover anything *new* about sex in reading the book, I can't see how this ends up interfering with a parent's right to decide when and where their kids discover the existence of sex.

Nature's Daredevils:
Writing for Young Audiences

I know, at the end of the last column I promised that this issue I'd talk all about Macropayments, but that was before I found out that this was *Locus*'s special young-adult SF issue, and as I happen to be on tour with my first young-adult SF novel, *Little Brother*, I think I'd better put off Macros for a month a talk a little about what I've learned about writing for young people.

First of all, YA SF is gigantic and invisible. The numbers speak for themselves: a YA bestseller is likely to be moving ten times as many copies as an adult SF title occupying the comparable slot on the grownup list. Like many commercially successful things, YA is largely ignored by the power brokers of the field, rarely showing up on the Hugo ballot (and when was the last time you went to a Golden Duck Award ceremony?). Yet so many of us came into the field through YA, and it's YA SF that will bring the next generation into the fold.

Genre YA fiction has an army of promoters outside of the field: teachers, librarians, and specialist booksellers are keenly aware of the difference the right book can make to the right kid at the right time, and they spend a lot of time trying to figure out how to convince kids to try out a book. Kids are naturals for this, since they really use books as markers of their social identity, so that good books sweep through their social circles like chickenpox epidemics, infecting their language and outlook on life. That's one of the most wonderful things about writing for younger audiences—it *matters*. We all read for entertainment, no matter how old we are, but kids also read to find out how the world works. They pay keen attention, they argue back. There's a consequentiality to writing for young people that makes it immensely satisfying. You see it when you run into them in person and find out that there are kids who read your book, googled every aspect of it, figured out how to replicate the best bits, and have turned your story into a hobby. We wring our hands a lot about the greying of SF, with good reason. Just have a look around at your regional con, the one you've been going to since you were a teenager, and count how many teenagers are there now. And yet, young people are reading in larger numbers than they have in recent memory. Part of that is surely down to Harry Potter, but on this tour, I've discovered that there's a legion of unsung heroes of the kids-lit revolution.

These are booksellers like Anderson's of Naperville, a suburb of Chicago. Anderson's operates a lovely bookstore, the kind of friendly indie shop that we all have cherished memories of, but that's not the main event. They *also* have a "book fair" business that is run out of a nearby warehouse. This involves filling trucks with clever, rolling bookcases that snap shut like a cigarette case, each one pre-stocked with carefully curated titles. These are schlepped out to schools across the Midwest, assembled in impromptu school-gym book fairs. This matters. It matters because you don't go to the bookstore until you already know you love books. You need a gateway drug to get you hooked on the harder stuff. Traditionally, this was the non-bookstore retailer, the pharmacy, the supermarket. That was before distributorship consolidation took place in the wake of the rise of national, big-box retailers like Wal-Mart. The contraction in distribution led to a massive reduction in the number of titles stocked outside of bookstores across the land. Even as the bookstores got bigger and more elaborate, the vital induction system of finding the right book on the right spinner rack at the right age was collapsing. We know what happened next: the collapse of the midlist, massive mergers and acquisitions in publishing, the shuttering of many longstanding careers in the field. So when Anderson's pulls up a truck outside your school and puts the books right there, where you can't miss them, it starts to take back that vital territory that we lost 20 years ago, starts to bring kids back into the faith.

Writing for young people is really exciting. As one YA writer told me, "Adolescence is a series of brave, irreversible decisions." One day, you're someone who's never told a lie of consequence; the next day you have, and you can never go back. One day, you're someone who's never done anything noble for a friend; the next day you have, and you can never go back. Is it any wonder that young people experience a camaraderie as intense as combat-buddies? Is it any wonder that the parts of our brain that govern risk-assessment don't fully develop until adulthood? Who would take such brave chances, such existential risks, if she or he had a fully functional risk-assessment system?

So young people live in a world characterized by intense drama, by choices wise and foolish and always brave. This is a book-plotter's dream. Once you realize that your characters are living in this state of heightened consequence, every plot-point acquires moment and import that keeps the pages turning.

The lack of regard for YA fiction in the mainstream isn't an altogether bad thing. There's something to be said for living in a disreputable, ghettoized bohemia (something that old-time SF and comics fans have a keen appreciation for). There's a lot of room for artistic, political, and commercial expectation over here in low-stakes land, the same way that there was so much room for experimentation in other ghettos, from hip-hop to roleplaying games to dime-novels. Sure, we're vulnerable to moral panics about corrupting youth (a phenomenon as old as Socrates, and a charge that has been leveled at everything from the waltz to the jukebox), but if you're upsetting that kind of person, you're probably doing something right.

Risk-taking behavior—including ill-advised social, sexual, and substance adventures—are characteristic of youth itself, so it's natural that anything that co-occurs with youth, like SF or TV or video games, will carry the blame for them. However, the frightened and easily offended are doing a better job than they ever have of collapsing the horizons of young people, denying them the pleasures of gathering in public or online for fear of meteor-strike-rare lurid pedophile bogeymen, or on the pretense of fighting gangs or school shootings or some other tabloid horror. Literature may be the last escape available to young people today. It's an honor to be writing for them.

Beyond Censorware: Teaching Web Literacy

The Problem

Control over the way kids use computers is a real political football, part of the wide-ranging debates over child pornography, bullying, sexual predation, privacy, piracy, and cheating.

And if those stakes weren't high enough, consider this: The norms of technology use that today's kids grow up with will play a key role in tomorrow's workplace, national competitiveness, and political discourse.

Hoo-boy—poor kids!

The idea that kids can run technological circles around their elders is hardly new. In 1878, the newly launched Bell System was crashed by its operators, young messenger boys who'd been redeployed to run the nascent phone system and instead treated the nation's fragile communications infrastructure as the raw material for a series of pranks and ill-conceived experiments.

Today, kids are still way ahead of the grownups who supposedly control their school and home networks. In my informal interviews, I've discovered again and again that kids are a bottomless well of tricks for evading network filters and controls, and that they propagate their tricks like crazy, trading them like bubble-gum cards and amassing social capital by helping their peers gain access to the whole wide Web, rather than the narrow slice that's visible through the crack in the firewall.

I have to admit, this warms my heart. After all, do we want to raise a generation of kids who have the tech savvy of an Iranian dissident, or the ham-fisted incompetence of the government those dissidents are running circles around?

But I'm also a parent, and I know that it won't be long before my daughter is using her network access to get at stuff that's so vile, my eyes water just thinking about it. What's more, she's going to be exposed to a vast panoply of privacy dangers, from the marketing creeps who'll track her around the Web to the spyware jerks who'll try to infect her machine to the crazed spooks at agencies like the NSA who are literally out to wiretap the entire world.

Add to that the possibility that the disclosures she makes on the network are likely to follow her for her whole life, every embarrassing utterance preserved for eternity, and it's clear that there's a problem here.

But I think I have the solution. Read on.

The Solution: No Censorware

Let's start by admitting that censorware doesn't work. It catches vast amounts of legitimate material, interfering with teachers' lesson planning and students' research alike.

Censorware also allows enormous amounts of bad stuff through, from malware to porn. There simply aren't enough prudes in the vast censorware boiler-rooms to accurately classify every document on the Web.

Worst of all, censorware teaches kids that the normal

course of online life involves being spied upon for every click, tweet, email, and IM.

These are the same kids who we're desperately trying to warn away from disclosing personal information and compromising photos on social networks. They understand that actions speak louder than words: If you wiretap every student in the school and punish those who try to get out from under the all-seeing eye, you're saying, "Privacy is worthless."

After you've done that, there's no amount of admonishments to value your privacy that can make up for it.

On the other hand, censorware provides a brilliant foil for a curriculum unit that teaches 21st century media literacy in ways that are meaningful, informative, and likely to make kids and the networks they use better and safer.

The Lesson Plan

Here's my outline for a curriculum of media literacy (addressed to the students):

> 1) Work with your teacher to select 30 important keywords relevant to your curriculum. Check the top 50 results for each on Google or another popular search engine, and record how many are blocked by your school firewall.
>
> A study undertaken by the Electronic Frontier Foundation in 2003 found that up to *50 percent* of pages relevant to common U.S. curricula were

blocked by various commercial censorware products.

In this exercise, students learn comparative searching, statistical analysis, and gain greater familiarity with their own curricula.

Further study includes identifying those subjects that are more apt to be blocked—for example, sites relevant to reproductive health, breast cancer, racism, etc.

2) Keep a log of the inappropriate pages you encounter while browsing, including pornographic pages, adware, malware, and so on. Compile a chart showing how many times a day your school's censorware fails to protect the students in your class.

More stats here, introducing the idea of both "false positive" and "false negative." This also opens the debate on what is and is not a "bad" page, demonstrating how subjective this kind of classification is.

3) Interview your teachers about the ways that censorware interferes with their teaching.

Every teacher has stories about cueing up a video in the morning to use after lunch, only to discover that it's been blocked in the interim and blown their lesson-plan. Gathering these stories helps students understand that censorware affects everyone in the school, even the teachers who are supposedly in charge of their care and education.

4) Interview your fellow students about the ways that they defeat censorware (e.g., looking for unblocked proxies by searching for "proxy" on Google and moving to the 75th page of results, so deep that it's unlikely to have been catalogued by the censorware companies; or evading blocks on message-boards by using random, ancient blog-posts' message areas to conduct secret conversations).

Discuss "security through obscurity" and "security theater," and whether a security system can be said to work if it can be so trivially evaded by kids.

5) Research the corporate reputation and practices of the censorware company that supplies your school.

Many censorware companies have very dirty hands. For example, SmartFilter (now a division of McAfee) is a high-profile supplier of censorware to repressive regimes. SmartFilter's software was recently used in the United Arab Emirates to block news about a member of the royal family who had been video-recorded brutally torturing a business-associate. Learning to research the credibility and conduct of people who provide information on the internet is the key to understanding which information can and cannot be trusted.

6) Contact the censorware company and ask for the

criteria by which it rates pages. Ask for the reason that the false positives identified in step 1 were classified as objectionable.

7) Research how to file a Freedom of Information Act (FOIA) request and use the procedure to discover how much your school or school board spends on censorware.

Imagine what kind of nation we'd have if every high-school graduate knew how to file an FOIA request—once you've learned this, no civics, history, or politics class will ever be the same.

8) Bonus marks: Present your research to your board of education.

Get on the agenda for an upcoming meeting. Present your findings: Our censorware fails to protect us in these ways; it interferes with our education in these ways; it is technically insufficient in these ways; the company is unworthy of public funds in these ways; the money could be spent on this, that, and the other. Thank you.

A unit like this, undertaken as a year-long project, would graduate a generation of students who understand applied statistics, risk and security, civic engagement, legal procedures, and the means by which you can evaluate the information you receive.

What more could any society ask for?

Writing in the Age of Distraction

We know that our readers are distracted and sometimes even overwhelmed by the myriad distractions that lie one click away on the internet, but of course writers face the same glorious problem: the delirious world of information and communication and community that lurks behind your screen, one alt-tab away from your word-processor.

The single worst piece of writing advice I ever got was to stay away from the internet because it would only waste my time and wouldn't help my writing. This advice was wrong creatively, professionally, artistically, and personally, but I know where the writer who doled it out was coming from. Every now and again, when I see a new website, game, or service, I sense the tug of an attention black hole: a time-sink that is just waiting to fill my every discretionary moment with distraction. As a co-parenting new father who writes at least a book per year, half-a-dozen columns a month, ten or more blog posts a day, plus assorted novellas and stories and speeches, I know just how short time can be and how dangerous distraction is.

But the internet has been very good to me. It's informed my creativity and aesthetics, it's benefited me professionally and personally, and for every moment it steals, it gives back a hundred delights. I'd no sooner give it up than I'd give up fiction or any other pleasurable vice.

I think I've managed to balance things out through a few simple techniques that I've been refining for years. I still sometimes feel frazzled and info-whelmed, but that's rare. Most of the time, I'm on top of my workload and my muse. Here's how I do it:

- Short, regular work schedule
  When I'm working on a story or novel, I set a modest daily goal—usually a page or two—and then I meet it every day, *doing nothing else* while I'm working on it. It's not plausible or desirable to try to get the world to go away for hours at a time, but it's entirely possible to make it all shut up for 20 minutes. Writing a page every day gets me more than a novel per year—do the math—and there's always 20 minutes to be found in a day, no matter what else is going on. Twenty minutes is a short enough interval that it can be claimed from a sleep or meal-break (though this shouldn't become a habit). The secret is to do it every day, weekends included, to keep the momentum going, and to allow your thoughts to wander to your next day's page between sessions. Try to find one or two vivid sensory details to work into the next page, or a *bon mot*, so that you've already got some material when you sit down at the keyboard.

- Leave yourself a rough edge
  When you hit your daily word-goal, **stop**. Stop even if you're in the middle of a sentence. Especially if you're in the middle of a sentence. That way, when you sit down at the keyboard the next day, your first five or

ten words are already ordained, so that you get a little push before you begin your work. Knitters leave a bit of yarn sticking out of the day's knitting so they know where to pick up the next day—they call it the "hint." Potters leave a rough edge on the wet clay before they wrap it in plastic for the night—it's hard to build on a smooth edge.

- Don't research

  Researching isn't writing and vice-versa. When you come to a factual matter that you could google in a matter of seconds, *don't*. Don't give in and look up the length of the Brooklyn Bridge, the population of Rhode Island, or the distance to the Sun. That way lies distraction—an endless click-trance that will turn your 20 minutes of composing into a half-day's idyll through the web. Instead, do what journalists do: type "TK" where your fact should go, as in "The Brooklyn Bridge, all TK feet of it, sailed into the air like a kite." "TK" appears in very few English words (the one I get tripped up on is "Atkins") so a quick search through your document for "TK" will tell you whether you have any fact-checking to do afterwards. And your editor and copyeditor will recognize it if you miss it and bring it to your attention.

- Don't be ceremonious

  Forget advice about finding the right atmosphere to coax your muse into the room. Forget candles, music, silence, a good chair, a cigarette, or putting the kids to sleep. It's nice to have all your physical needs met before you write, but if you convince yourself that you can only write in a perfect world, you compound the problem of finding 20 free minutes with the problem of finding the right environment at the same time. When the time is available, just put fingers to keyboard and write. You can put up with noise/silence/kids/discomfort/hunger for 20 minutes.

- Kill your word-processor
  Word, Google Office, and OpenOffice all come with a bewildering array of typesetting and automation settings that you can play with forever. Forget it. All that stuff is distraction, and the last thing you want is your tool second-guessing you, "correcting" your spelling, criticizing your sentence structure, and so on. The programmers who wrote your word processor type all day long, every day, and they have the power to buy or acquire any tool they can imagine for entering text into a computer. They don't write their software with Word. They use a text-editor, like vi, Emacs, TextPad, BBEdit, Gedit, or any of a host of editors. These are some of the most venerable, reliable, powerful tools in the history of software (since they're at the core of all other software) and they have almost no distracting features—but they do have powerful search-and-replace functions. Best of all, the humble .txt file can be read by practically every application on your computer, can be pasted directly into an email, and can't transmit a virus.

- Realtime communications tools are deadly
  The biggest impediment to concentration is your computer's ecosystem of interruption technologies: IM, email alerts, RSS alerts, Skype rings, etc. Anything that requires you to wait for a response, even subconsciously, occupies your attention. Anything that leaps up on your screen to announce something new occupies your attention. The more you can train your friends and family to use email, message boards, and similar technologies that allow you to save up your conversation for planned sessions instead of demanding your attention *right now* helps you carve out your 20 minutes. By all means, schedule a chat—voice, text, or video—when it's needed, but leaving your IM running is like sitting down to work after hanging a giant "DISTRACT ME" sign over your desk, one that shines brightly enough to be seen by the entire world.

I don't claim to have invented these techniques, but they're the ones that have made the 21st century a good one for me.

Extreme Geek

I am by no means the geekiest SF writer working in the field today; on the power-law curve of geekiness, there are many ancient and gnarly masters before whom I am but a novitiate, barely qualified to check the syntax in their shell-scripts. Stross, I'm looking at you here.

Nevertheless, I am far *more* geeky than average, and that geekiness has crept into my writing practice in a way that is very close to perfectly geeky inasmuch as it probably costs me as much effort as it saves me, inasmuch as it delights me, and inasmuch as it points the way to civilian applications that someone else might want to develop into products that the less geekified may enjoy.

In that spirit, I offer you three quirky little tassles from the fringes of technology and SF writing:

1. Business: Book donation program

This is the lowest-tech entry on the list, but it's also the most generally applicable. As you know (Bob), I give away all my books as free, Creative Commons–licensed e-books the same day they go on sale in stores, on the grounds that for most people, a free e-book is more apt to entice them to buy the print book than to substitute for it.

But there's a small minority—mostly other geeks—for whom the e-book is *all* they want, and who, nevertheless, want to see the writers they enjoy compensated (bless 'em!). They write to me with some variation on, "Can't I just send you a donation?" And my answer has always been no, because:

1. I don't want to have to bookkeep, file taxes on, and otherwise track your $5;

2. I don't want to cut my extremely valuable and useful publisher out of the loop;

3. I don't want to reduce my print-books' sell-through rates (which determine advance sizes, print runs, and bookstore orders).

So, traditionally, I asked my readers to compensate me by donating a book to a school or library or halfway house. But, practically speaking, this isn't very useful advice. Most of us have no idea how to give books away to schools or libraries—do you just show up at the reception desk with a book, shove it into the clerk's hands and say, "Here, this is for you?"

Starting with my novel *Little Brother*, I've been doing something different: I actually provide a matchmaking service to connect donors with willing recipients. I hired an assistant—the talented Olga Nunes—to monitor through a googlemail address that I published in a solicitation to schools, libraries, etc., telling them to email their work contact details if they wanted a free copy of the book. Olga vetted these to ensure that they weren't fakers or scam artists, and then posted a geographically sorted list of would-be donees to my site.

Then, I put the word out to potential donors that there was an easy (or at least *easier*) way to compensate me if you liked the e-book and didn't need the hardcopy: visit your favorite bookstore and buy as many copies as you'd like for any of the organizations that solicited donations, then email us the receipt so we can cross them off the list. Judging from donor emails, many of them just gave to the first outstanding request, others looked for requests from their region, and others judged by merit. Some donated several copies—as many as 15! As I type this, we've given away well over 200 copies to people who really wanted the book. I got the sales number, my publisher got the sale, the library or school got the material, and the reader got to feel like s/he had paid for the value s/he'd received.

Now, this wasn't cheap. I needed to hire someone with the good judgment to tell scammers from honest people and with the HTML skills to format and update the page. I definitely spent at least twice as much as I made on this program. As a commercial venture, it was a flop.

But as a proof-of-concept, it was a ringing success. There *is* a market opportunity here for someone who wants to automate the service. I envision something run jointly by, say, the American Library Association (or maybe the International Federation of Library Associations) and the Adopt-a-School program (to ease vetting), that works with a couple dozen booksellers, national and local, and lists books by all kinds of authors and requests from all over the world. Donors can either get a suggestion for a book to donate (perhaps based on preferences like "Science Fiction" or "Young-Adult Novels" and "Schools in My Area" or "Schools in the Nation's Poorest ZIP Codes") and, with a few clicks, donate a book, receiving a tax-deduction receipt in return.

2. Research: Twitter meets notekeeping

I'm in the middle of a research-intensive novel, for which I've read some 50 or 60 books. I made extensive notes as I did, unconsciously falling into a Twitter-style shorthand in my long text-file, for example:

> • Newborn babies are swaddled tightly at birth, it tames them. If you aren't swaddled, you grow up wild and restless. Socialism 79 #china #childhood #control
>
> • Louche boy wearing wide-bottom "trumpet trousers" and shirt rolled up to expose his belly on a hot day. Socialism 86 #china #fashion
>
> • "Drink vinegar" is "conjugal jealousy." Socialism 155 #china #slang #romance

These notes are from *Socialism Is Great!*, Lijia Zhang's amazing memoir of life in rural China during the period of economic reform and industrialization. The hashtags (#tag) are loose categories that each note seemed to fit into while I was writing them down. These notes, and hundreds more, live in a text file.

As I made these notes, I had a sense that, somewhere, there'd be a program that would parse through them, generating a tag-cloud [see picture] with clickable links to different hashtags' contents. Unfortunately, as this file grew longer, I realized that no such program existed.

adolescence age atrocity authoritarianism banking bargaining betrayal birthcontrol bollywood booms boss
bureaucracy business chauvinism childhood children china chinatown city classwar coase
colonialism commerce commodities communism conflict consumerism contract control cops corruption
crime culture culture] death deception democracy developingnations development diaspora disparity dorm driving
economics education emotion employment entrepreneurship environment factory failure fakes
falungong family fashion fighting film films food forgery frugality funny game gamer games gaming gangs
gangsters gender generationgap goldfarm goldfarming gore groups gun hierarchy hinglish history
hoarse hygeine identitytheft immigration india infringing injury internationalism jail justice labor
language law leisure lifestyle literacy love magic maker manfuacturing manufacturing mao medicine migration
mods money moral motorcycles music negotiation opium organizing papers parents phone pinkertons
players plot police pollution ponzi population poverty prison privacy profanity protest proverb quants race racism
rain recycling religion reversal revolution ritual romance safety savings scab school selfhelp sex sez shoes
slang slums smuggling socialmobility solidarity song songs speech spies stats status streetlife strike
strikes success suicide superstition surveillance sweatshop technology trade union unions usa violence voting
weapon weather weird wto youth

Search all notes for: Search
(search is case-insensitive)

I put the call out to the readership at Boing Boing, the blog I co-edit, and Dan McDonald, one of my readers, came through with a fantastic little Perl script called tagcloud.pl that does exactly this, parsing all my notes into a database that I can search or query visually, by clicking on the cloud.

Now, as I write the novel, this has become an invaluable aid: for one thing, it lends itself to a kind of casual, clicky browsing in which one hashtag leads to another, to a search-query, to another tag, exploring my notes in a way that is both serendipitous and directed.

For another, the format is one that comes naturally to me, because of all the other services I use—such as Twitter—that employ this telegraphic, brief style.

Dan's Perl script is freely licensed and can be downloaded from perlmonks.org/?node_id=707360.

3. Process: Flashbake

I know a lot of archivists and one of their most common laments is the disappearance of the distinct draft manuscript in the digital age. Pre-digital, authors would create a series of drafts for their work, often bearing hand-written notations tracking the thinking behind each revision. By comparing these drafts, archivists and scholars could glean insights into the author's mental state and creative process.

But in the digital era, many authors work from a single file, modifying it incrementally for each revision. There are no distinct, individual drafts, merely an eternally changing scroll that is forever in flux. When the book is finished, all the intermediate steps that the manuscript went through disappear.

It occurred to me that there was no reason that this had to be so. Computers can remember an insane amount of information about the modification history of files—indeed, that's the norm in software development, where code repositories are used to keep track of each change to the codebase, noting who made the changes, what s/he changed, and any notes s/he made about the reason for the change.

So I wrote to a programmer friend of mine, Thomas Gideon, who hosts the excellent Command Line podcast (http://thecommandline.net), and asked him which version control system he'd recommend for my fiction projects—which one would be easiest to automate so that every couple of minutes, it checked to see if any of the master files for my novels had been updated, and then check the updated ones in.

Thomas loved the idea and ran with it, creating a script that made use of the free and open-source control system "Git" (the system used to maintain the Linux kernel), checking in my prose at 15-minute intervals, noting, with each check-in, the current time-zone on my system clock (where am I?), the weather there, as fetched from Google (what's it like?), and the headlines from my last three Boing Boing posts (what am I thinking?). Future versions will support plug-ins to capture even richer metadata—say, the last three tweets I twittered, and the last three songs my music player played for me.

He called it "Flashbake," a neologism from my first novel, *Down and Out in the Magic Kingdom*. I was honored.

It's an incredibly rich—even narcissistic—amount of detail to capture about the writing process, but there's no reason *not* to capture it. It doesn't cost any more to capture all this stuff every 15 minutes than it would to capture a daily file-change snapshot at midnight without any additional detail. And since Git—and other source repositories—is designed to let you summarize many changes at a time (say, all the changes between version 1 and version 2 of a product), it's easy to ignore the metadata if it's getting in the way.

Now, this may be of use to some notional scholar who wants to study my work in a hundred years, but I'm more interested in the immediate uses I'll be able to put it to—for example, summarizing all the typos I've caught and corrected between printings of my books. Flashbake also means that I'm extremely backed up (Git is designed to replicate its database to other servers, in order to allow multiple programmers to work on the same file). And more importantly, I'm keen to see what insights this brings to light for me about my own process. I know that there are days when the prose really flows, and there are days when I have to squeeze out each word. What I don't know is what external factors may bear on this.

In a year, or two, or three, I'll be able to use the Flashbake to generate some really interesting charts and stats about how I write: does the weather matter? Do I write more when I'm blogging more? Do "fast" writing days come in a cycle? Do I write faster on the road or at home? I know myself well enough to understand that if I don't write down these observations and become an empiricist of my own life that all I'll get are impressionistic memories that are more apt to reflect back my own conclusions to me than to inform me of things I haven't noticed.

Thomas has released Flashbake as free/open software. You can download it and start tinkering at http://bitbucketlabs.net/flashbake. As I said, it's not the kind of thing that an info-civilian will be able to get using without a lot of tinkering, but in the month I've used it, I've already found it to be endlessly fascinating and useful—and with enough interest, it's bound to get easier and easier.

How to Stop Your Inbox Exploding

I live and die in my email, receiving hundreds of non-spam messages every day. If I'm stationary and not actually feeding or playing with the baby, chances are I've got my laptop open somewhere nearby, online and downloading mail. It's my alpha and my omega, my version control system (if I want to find an old version of a document, I just find the copy I emailed to someone earlier), my address book, my journal, and my confessor. I have over a million archived pieces of email, going back to 1991.

What's more, my inbox is almost always *empty*.

I've spent more than a decade tinkering with my email workflow, perfecting it so that I can manage whatever life throws at me through my inbox. I've come up with a few tips and hacks that never fail to surprise and delight my friends and colleagues when I show them off, and here they are, for the record:

Sort your inbox by subject

This is my favorite one by far. If something big is going on in the world, chances are lots of people are going to be emailing you about it, and they'll generally use pretty similar subject lines.

When my daughter was born, the majority of congratulatory emails began with the word "Congratulations." When I'd asked my friends to help me find an office, most of the tips I got began with "office."

Best of all, if some spammer manages to get a few hundred copies of a message through my filter and into my inbox, they'll all have the same subject line, making them easy to bulk-select and delete.

Foreign-alphabet spam is also a doddle, since non-Roman characters will all alphabetise at the bottom or top of your inbox; if you don't read Cyrillic, Korean, Hebrew, or Simplified Kanji, you can just delete them all with a couple of key presses.

Colour-code messages from known senders

Somewhere in the guts of your email client is a simple tool for adding "rules" or "filters" for the mail you send and receive. Here's a simple pair that have made my mail more manageable: first, add to your address book everyone who receives mail from you; second, change the colour of messages from known senders to a different tone from your regular mail (I use a soothing green).

This lets you tell, at a glance, whether a message is from someone you've seen fit to send a message to in times gone by. This is particularly useful for picking misidentified spam out of your spam folder: anything from a known sender that your mailer mistakenly stuck in there is probably worth a closer look.

Kill people who make you crazy

If there's someone—often a stranger—who has found you via the internet and taken it upon her/himself to make your life a living hell by sending you pointless, argumentative messages, don't rely on your own iron discipline to keep from reading and responding to these mean little darts.

Instead, create a filter called "killfile" and add the email addresses of these anti-correspondents to it, then instruct your mailer to either delete or tuck away these messages somewhere you won't have to look at them.

Half-resign from mailing lists

Many of us are obliged to join up to mailing lists for social or work or family reasons, even though most of the messages are irrelevant to us—for example, a list for planning an annual event in which you play some small part. Resigning from these lists isn't an option, but you can't read them all, either. A nice middle ground is to write a mail-rule that files all mailing list messages in a "mailing list" folder, *except* for those individual messages that contain your name ("If subject line contains [name of list] and body does not contain [your name] then…"). That way, you'll be able to respond immediately whenever someone brings you up in conversation, and the rest of the messages will be safely tucked away for you to refer to.

Keep a pending list

I have a little text-file on my desktop called WAITING. TXT that lists every call, email, parcel, and payment I'm waiting for, in simple form (e.g. WAITING EMAIL Fred on dinner 1/5/8 or WAITING MONEY £32.11 Refund from John Lewis). Once or twice a day, I cast my eye over the list and see if there's anything that I should have heard on but haven't, and send a reminder. This has saved more dinner dates, money, and time than anything else on the list.

What I Do

From time to time, people ask me for an inventory of the tools and systems I use to get my work done. As a hard-traveling, working writer, I spend a lot of time tinkering with my tools and systems. At the risk of descending into self-indulgence (every columnist's occasional privilege), I'm going to try to create a brief inventory, along with a wish/to-do list for the next round.

First, the hardware:

Laptop: ThinkPad X200. This is the next-to-most-recent version of Lenovo's ThinkPad X-series, their lightweight travel notebooks. The X200 is fast enough that it never feels slow, and like all ThinkPads, is remarkably rugged and easy to do small maintenance chores upon. I bought mine in the UK but I prefer a U.S. keyboard; I ordered one of these separately and did the swap in 20 seconds flat without ever having done it before. I bought my own 500GB hard drive and 4GB of RAM separately (manufacturers always gouge on hard drives and memory) and installed them in about five minutes. Lenovo bought the ThinkPad line from IBM in 2005, but IBM still has the maintenance contract, through its IBM Global Services division. For $100 or so a year, I was able to buy an on-site/next-day hardware replacement warranty that means that when anything goes wrong with the laptop hardware, IBM sends a technician out to me the next day, with all the necessary replacement parts, no matter where I am in the world.

I've been using ThinkPads since 2006 and have had occasion to use this maintenance contract three times, and all three times I was favorably impressed (lest you think three servicings in four years is an indicator of poor hardware quality, consider that every other brand of computer I've carried for any length of time became fatally wounded in less than a year).

I have two different batteries for the ThinkPad: a 9-cell that weighs 1.4 pounds, and a little 4-cell that weighs a mere 10 oz. I only use the 9-cell while traveling: it's good for 5+ hours, even while powering a wireless link. When I'm at home, I spare myself the additional weight by switching to the 4-cell, which makes my daily walk to and from the office a lot more comfortable for my aching lumbar. The 4-cell's only good for an hour or two, but I'm rarely away from a power-outlet for longer than that while at home.

I have a dock for the ThinkPad at the office, connected to a generic Sanyo monitor and a truly stupendous Datamancer hand-made steampunk keyboard (http://datamancer.net—go look. Drool. Then come back). I also have a Logitech Anywhere MX mouse, which is the first mouse I've used in years that really excited me: very precise, great ergonomics, and a wheel that you can unclutch so that it spins freely, making it easy to get to the bottom of a long file. It's also very satisfying, a little *whee* every time you send it whizzing. The dock also has a DVD/CD drive (the superportable ThinkPad models don't—I don't miss it).

I also have a backup drive at the office: just a generic full-height 500GB drive that was cheap on Amazon. I also have a little USB-powered generic 500GB 9mm SATA drive that I travel with. When I'm at home, I backup to the full-size drive every day when I sit down at my desk; on the road, I run the backup over breakfast.

Wish list: Lenovo's just shipped the X201: faster, with a touchpad. Want. Don't need it in any meaningful way, but I am Pavlovian about upgrade paths. Can't wait for the fast, low-power-draw solid state drives to get up to 500GB, the minimum size for my needs.

Phone: I've got a Google/HTC NexusOne, and like Tim Berners-Lee, I can solemnly declare that I hate this phone less than any other phone I've ever owned. I rooted the phone, following simple instructions I found online, and now I can use it as a modem for my laptop, which is insanely awesome, especially on book-tours and at conferences where 4,000 geeks have saturated the hotel's net connection. I have SIMs with unlimited data-plans for T-Mobile U.S. and Orange UK, and switch depending on which country I'm in. The NexusOne also comes with turn-by-turn satellite navigation and a Google Maps app that factors in local traffic data. As such, it has enabled me to save $10–$15/day when traveling by omitting the GPS for the rental car. Unlike its predecessor, the G1, the NexusOne is fast enough to run the stellar Android operating system.

Now, onto the local software:

Operating system: I'm using Ubuntu, a version of the free and open GNU/Linux operating system that is designed to be easy to use and maintain for non-techy people. I was once a Unix sysadmin, but it has been a long time, and I wouldn't hire me to do it today. Ubuntu Just

Works. I recently had cause to install Windows XP on an old ThinkPad and found that it was about a hundred times more complicated than getting Ubuntu running. When I transitioned to Ubuntu from the MacOS, I had a week or two's worth of disorientation, similar to what happened after we renovated the kitchen and changed where we kept everything. Then the OS just disappeared, and it has stayed disappeared, breaking in ways that are neither more severe nor more frequent than any other OS I've ever used.

The only times Ubuntu asserts itself as an artifact (rather than as invisible plumbing) is when it is impressing me with spectacular Just Workingness. For example, Ubuntu's facility for finding and installing apps easily also means that when you migrate to a new computer (something I do every 8–12 months), you can just feed the new Ubuntu installation an automatically generated list of all the programs you run and it will quietly and efficiently install all of them and get them configured. Another example: Ubuntu's support for 3G wireless modems is vastly superior to the experience on the Mac and under Windows, where the 3G drivers are commercial and typically supplied by the cellular companies. These proprietary drivers come with all kinds of crapware and like to throw up big splashy screens announcing that you have connected to the internet (with an implied string of exclamation points: !!!!!!!), which gets old the bazillionth time you plug in the modem to get directions or check email. By contrast, Ubuntu uses its own, pre-installed drivers that Just Work: plug in a modem, and it asks you which country you're in and which carrier you're on. Then it sets up the modem, perfectly, every time I've tried it. It's astounding.

Email: I live in email. It's probably a generational thing, but I can't understand the received wisdom that the next generation of computer users prefers IM to email. For me, email is a powerful organizing force, a 1.5 million–piece archive that represents my entire professional and personal history. Old versions of stories, letters to friends, commercial orders—if it's not in my email archive, I don't know it.

I use Thunderbird, an industrial-strength local email client that's free and open, overseen by the Mozilla Foundation, best known for their Firefox browser. I find the spam filtering tolerably good, and I augment it by automatically adding every email address I reply to to my address book, then using a filter to automatically color email from my past correspondents green, so that I can see at a glance if there's anything in my junk folder from someone I've previously traded emails with. I also color emails blue if they're from strangers where I'm the "To" address (as opposed to a CC), which is a good way of quickly spotting personal emails as opposed to spam from PR people.

I store my archived emails in nested folders: Friends (2010, 2009, 2008...), Boing Boing (2010, 2009, 2008...), Commerce (2010, etc.), Speaking Gigs/Travel, Activism, Writing (General, then subfolders for each book and each magazine that I regularly write for). I file both sent and received emails, doing it routinely through the day.

I access my email through an SSH tunnel, which is handy for contexts where the networks block access to my SMTP server, and it keeps my messages private from local snoops. There's a cron job (a regularly scheduled task) that checks every minute or two to make sure the tunnel is up, and if it isn't, it restarts it (because I lose the tunnel every time I change network connections).

On the mobile side, I use an open/free Android POP client called K-9 mail. It's a little primitive—it could use better filtering and status indicators, and it's a huge pain in the ass to undelete an accidentally deleted message with. But it's OK. I have it set to POP my server but not delete messages unless I delete them on the device too. My pattern is to use the phone to check (but usually not to reply to) mail between laptop sessions. I delete anything dumb or spammy (so I don't have to delete it again on the laptop). K-9 is smart enough to clear local caches of the messages once they've been downloaded to the laptop and deleted from the server.

Wishlist: I dream of a faster, more robust search for Thunderbird. I have so much useful and important info in my archived email, but Thunderbird is slow and poky when it comes to searching through all those millions of messages. I also miss the days when I ran my own local IMAP server on my laptop and used several email clients to access it, which let me use one client for spam-filtering, another for blog-business, another for search, and another for reading, while IMAP kept it all in synch. I gave it up because all those multiple copies of my email corpus overflowed my hard drive. I've got 8GB of email archives now, and keeping 4 or 5 copies of them would probably be more do-able than it was several years ago.

Browser: I use Firefox, along with a small group of

very useful plugins: CustomizeGoogle, which lets me see more search results (100 at a time), with miniature thumbnails for each; Linky (which lets me open a lot of links at once in multiple tabs, useful for articles that have been divided into multiple sections); and TinEye, an image search tool that helps me find the original version of images that I've located in anonymous corners of the web (great for making sure I credit the right source in a blog post). I also live and die by TabMix Plus, which gives me much more control over my tabs, including the vital "Unclose tab" function that lets me re-open a tab that I closed in haste. I like systems with forgiveness in them—they're much more human than systems that expect inhuman perfection.

I have a couple hundred sites in a folder that I open as a series of tabs a couple times a day, quickly zipping through them after they've loaded to see if anything new has been posted.

Calendar: Thunderbird again. I love electronic calendars and my database of appointments goes back to the early 1990s (very handy for looking up that restaurant I loved that time I was in Baltimore). I have yet to find a good way to synchronize my calendar with a mobile device, mostly because every calendar vendor has decided that all calendar entries should be time-zone-dependent, so if you're in London when you key in a 5PM flight, the computer "helpfully" switches that to 12PM when you change the system clock to New York time. I've got Thunderbird's calendar set to keep all its times in London time no matter where I am, but as soon as I synch it with a mobile device, the device tries to reset all my times.

Wishlist: I want a simple way to share calendars without worrying about timezones—if an item says Cory's on a 1200h flight," the person I'm sharing with should be able to know, with total certainty, that the "1200h" means "1200h in the timezone Cory is in." This would make coordinating with my wife, my publishers' publicists, and my travel assistant vastly simpler. Dammit.

RSS: I use Liferea, which isn't a great reader, but is OK. It's a lot faster than it used to be, but it has a slow, nonfunctional search and has no way to go back to the last item I zipped past too quickly. I've got a couple thousand RSS feeds, but I don't try to read them all, just whir past them skimming for interesting keywords.

Wishlist: I dream of having an RSS reader that will archive everything in every RSS feed I've ever read, and let me search it, fast, on my own hard drive. ZOMG. Drool. All that personalized corpus, in hyperlinked, cached, high-availability low-latency glory.

Office suite: I use the free/open OpenOffice.org. Mostly I use the spreadsheet program to keep my personal books, using linked spreadsheets I've been tweaking since I first incorporated in the early 1990s. My accountant likes them so well that she often asks if she can share blank sets with her other clients. I enter receipts daily, and go through the activity on my bank account every morning and check for anomalies.

I sometimes use the OOo word-processor, usually to do light formatting or for business correspondence.

Writing: I use a plain-jane text editor that comes

with Ubuntu called Gedit. It doesn't do anything except accept text and save it and let me search and replace it. There were a few text-wrangling features in BBEdit on the Mac that I miss, but not very much. I like writing in simple environments that don't do anything except remember what words I've thought up. It helps me resist the temptation to tinker with formatting. I also use Gedit to compose blog-posts for Boing Boing, typing the HTML by hand, which is an old habit from the early 1990s. I do use syntax coloring to help me spot unclosed tags, but apart from that, I don't use any automated tools.

Scripts: I have a few small utility scripts that I run from the command line as part of my daily life: there's a backup script that uses rsync (a secure and smart free incremental backup script), and another rsync script that uses ImageMagick (a free image manipulation library) to resize and upload images that I've saved to the desktop. A reader created a Firefox bookmarklet for me that automatically creates an attribution link to Flickr pages, which is useful when crediting Creative Commons licensed art I've pulled for use on Boing Boing.

I've written here before about Flashbake, the version-tracking program that Thomas Gideon created—it saves a snapshot of all my writing work every 15 minutes, along with the last three songs I've played, the last three posts I put on Boing Boing, my current location and timezone, and a few other environmental factors.

Other: I have a few other pieces of habitual utility software. I use the GIMP for image manipulation; digiKam for image organizing and Flickr uploading; Ksnapshot for sophisticated screenshotting, Banshee as a media player, VLC for videos (I sometimes put a small VLC window in one corner of my screen with my daughter's cartoons and she'll sit on my lap and watch while I do email or blogging, and we can each point to interesting things in each others' windows and talk about them, which is golden).

Finally, online services:

My personal blogs all run on WordPress, and I pay Mike Little, a freelance WordPress administrator, to do a little tinkering here and there with them. Recently, we installed eShop, which lets me sell my Random House Audio audiobooks as MP3s directly from the web. It's clunky but it gets the job done and it's free, and everything else was clunky and expensive.

Boing Boing runs on a very customized Movable Type, supported by one full-time managing editor, a part time sysadmin, and a contract programmer.

I use The Pirate Bay's IPREDator proxy service, which costs €5/month and is unlimited: by sending all my traffic through IPREDator's servers, I encrypt it in such a way that local snoops can't read it. IPREDator was created in response to Sweden's Draconian internet surveillance law (IPRED—the Intellectual Property Rights Enforcement Directive), which imposes a duty on ISPs to spy on their users in case they're infringing copyright. IPREDator stores no logs, and moves all traffic to the much-more-privacy-respecting Denmark before passing it on.

There are plenty of little bits and pieces I'm omitting—Seesmic for Twitter on Android; the cheapie Brookstone portable battery charger I'm trying out on this trip, the Ubuntu bootable maintenance USB stick I've got in my bag. But getting into every single little finicky detail would fill a book and go from self-indulgent to soporific, so I'll wrap it up here.

When I'm Dead, How Will My Loved Ones Break My Password?

Tales from the encrypt: If you care about the integrity of your data, it's time to investigate solutions for accessing and securing it—and not just for the here and now

Fatherhood changed me—for the better. It made me start to think on longer timescales, to ponder contingencies and contingencies for contingencies. My wife, too. Now that our daughter, Poesy, is 16 months old, we're settled in enough to begin pondering the imponderable.

We're ready to draw up our wills.

It was all pretty simple at first: it was easy to say what we'd do if one of us died without the other (pass assets on to the survivor), or if we both went together (it all goes in trust for the kid); even all three of us (a charity gets the money). I thought about a literary executor (don't want to risk my literary estate being inherited by a child who may grow up to be as weird about her father's creations as some of the notable litigious cranks who have inherited other literary estates), and found a writer I like and trust who agreed to take things on should I snuff it before the kid's gotten old enough for me to know I can trust her not to be a nut job about it all. We found a lawyer through a referral from another lawyer who'd already done great work for me—I checked him out and he seemed fine.

Then we hit a wall.

What about our data?

More specifically, what about the secrets that protect

our data? Like an increasing number of people who care about the security and integrity of their data, I have encrypted all my hard drives—the ones in my laptops and the backup drives, using 128-bit AES—the Advanced Encryption Standard. Without the passphrase that unlocks my key, the data on those drives is unrecoverable, barring major, seismic advances in quantum computing, or a fundamental revolution in computing. Once your data is cryptographically secured, all the computers on earth, working in unison, could not recover it on anything less than a geological timescale.

This is great news, of course. It means that I don't have to worry about being mugged for my laptop, or having my office burgled (even the critical-files backup I keep on Amazon S3's remote storage facility is guarded by industrial-strength crypto, so I'm immune from someone raiding Amazon's servers). The passphrase itself, a very long, complicated string, is only in my head, and I've never written it down.

This is fantastically liberating. I'm able to lock it all up: my private journals, my financial details, 15 years' worth of personal and professional correspondence, every word I've written since the early 1980s, every secret thought, unfinished idea, and work in progress. In theory, I could limit this cryptographic protection to a few key files, but that's vastly more complex than just locking up the whole shebang, and I don't have to worry that I've forgotten to lock up something that turns out to matter to me in 10 or 20 years.

I can even lock up all my passwords for everything else: email, banking, government services, social networking services, and so on, keeping them in a master file that is itself guarded by crypto on my drive.

So far, so good.

But what if I were killed or incapacitated before I managed to hand the passphrase over to an executor or solicitor who could use them to unlock all this stuff that will be critical to winding down my affairs—or keeping them going, in the event that I'm incapacitated? I don't want to simply hand the passphrase over to my wife, or my lawyer. Partly that's because the secrecy of a passphrase known only to one person and never written down is vastly superior to the secrecy of a passphrase that has been written down and stored in more than one place. Further, many countries' laws make it difficult or impossible for a court to order you to turn over your keys; once the passphrase is known by a third party, its security from legal attack is greatly undermined, as the law generally protects your knowledge of someone else's keys to a lesser extent than it protects your own.

I discarded any solution based on putting my keys in trust with a service that sends out an email unless you tell it not to every week—these "dead man's switch" services are far less deserving of my trust than, say, my wife or my solicitor.

I rejected a safe-deposit box because of all the horror stories I've heard of banks that refuse to allow access to boxes until the will is probated, and the data necessary to probate the will is in the box.

I pondered using something called Shamir's Secret Sharing Scheme (SSSS), a fiendishly clever crypto scheme that allows you to split a key into several pieces, in such a way that only a few of those pieces are needed to unlock the data. For example, you might split the key into 10 pieces and give them to 10 people such that any five of them can pool their pieces and gain access to your crypto-protected data. But I rejected this, too—too complicated to explain to civilians, and what's more, if the key could be recovered by five people getting together, I now had to trust that no five out of 10 people would act in concert against me. And I'd have to keep track of those 10 people for the rest of my life, ensuring that the key is always in a position to be recovered. Too many moving parts—literally.

Finally, I hit on a simple solution: I'd split the passphrase in two, and give half of it to my wife, and the other half to my parents' lawyer in Toronto. The lawyer is out of reach of a British court order, and my wife's half of the passphrase is useless without the lawyer's half (and she's out of reach of a Canadian court order). If a situation arises that demands that my lawyer get his half to my wife, he can dictate it over the phone, or encrypt it with her public key and email it to her, or just fly to London and give it to her.

As simple as this solution is, it leaves a few loose ends: first, what does my wife do to safeguard her half of the key should she perish with me? The answer is to entrust it to a second attorney in the UK (I can return the favour by sending her key to my lawyer in Toronto). Next, how do I transmit the key to the lawyer? I've opted for a written sheet of instructions, including the key, that I will print on my next visit to Canada and physically deliver to the lawyer.

What I found surprising all through this process was the lack of any kind of standard process for managing key escrow as part of estate planning. Military-grade crypto has been in civilian hands for decades now, and yet every lawyer I spoke to about this was baffled (and the cypherpunks I spoke to were baffling—given to insanely complex schemes that suggested to me that their executors were going to be spending months unwinding their keys before they could get on with the business of their estates, and woe betide their survivors, who'd be left in the cold while all this was taking place).

Meanwhile, I'm left with this conclusion: if you're not encrypting your data, you should be. And if you *are* encrypting your data, you need to figure this stuff out, before you get hit by a bus and doom your digital life to crypto oblivion.

Radical Presentism

Every writer has a FAQ—Frequently Awkward Question—or two, and for me, it's this one: "How is it possible to work as a science fiction writer, predicting the future, when everything is changing so quickly? Aren't you afraid that actual events will overtake the events you've described?"

It's a fresh-scrubbed, earnest kind of question, and the asker pays the compliment of casting you as Wise Prognosticator in the bargain, but I think it's junk. Science fiction writers don't predict the future (except accidentally), but if they're very good, they *may* manage to predict the present.

Mary Shelley wasn't worried about reanimated corpses stalking Europe, but by casting a technological innovation in the starring role of *Frankenstein*, she was able to tap into present-day fears about technology overpowering its masters and the hubris of the inventor. Orwell didn't worry about a future dominated by the view-screens from *1984*, he worried about a *present* in which technology was changing the balance of power, creating opportunities for the state to enforce its power over individuals at ever-more-granular levels.

Now, it's true that some writers will tell you they're extrapolating a future based on rigor and science, but they're just wrong. Karel Čapek coined the word *robot* to talk about the automation and dehumanization of the workplace. Asimov's robots were not supposed to be metaphors, but they sure acted like them, revealing the great writer's belief in a world where careful regulation could create positive outcomes for society. (How else to explain his idea that *all* robots would comply with the "three laws" for thousands of years? Or, in the Foundation series, the existence of a secret society that knows exactly how to exert its leverage to steer the course of human civilization for millennia?)

For some years now, science fiction has been in the grips of a conceit called the "Singularity"—the moment at which human and machine intelligence merge, creating a break with history beyond which the future cannot be predicted, because the post-humans who live there will be utterly unrecognizable to us in their emotions and motivations. Read one way, it's a sober prediction of the curve of history spiking infinity-ward in the near future (and many futurists will solemnly assure you that this is the case); read another way, it's just the anxiety of a generation of winners in the technology wars, now confronted by a new generation whose fluidity with technology is so awe-inspiring that it appears we have been out-evolved by our own progeny.

Science fiction writers who claim to be writing the future are more apt to be hamstrung by their timidity than by the pace of events. An old saw in science fiction is that a sci-fi writer can take the automobile and the movie theater and predict the drive-in. But the drive-in is dead, and the echoes of its social consequences are fading to negligibility; on the other hand, the fact that the automobile was responsible for the first form of widely carried photo ID and is thus the progenitor of the entire surveillance state went unremarked-upon by "predictive" sci-fi. Some of my favorite contemporary speculative fiction is instead nakedly allegorical in its approach to the future—or the past, as the case may be.

Consider Bruce Sterling's *The Caryatids* (Bantam, 2009), an environmental techno-thriller—Sterling once defined a techno-thriller as "A science fiction novel with the president in it"—set in a mid-twenty-first century in which global warming has done its catastrophic best to end humanity. Finally forced to confront the reality of anthropogenic climate change, humanity fizzles and factions off into three warring camps: the Dispensation, an Al-Gorean green-capitalist technocracy; the Acquis, libertarian technocrats who'll beta-test anything (preferably on themselves); and China, a technocracy based on the idea that technology can make command-and-control systems actually *work,* in contrast to the gigantic market failure that destroyed the planet. The play of these three ideologies serves as a brilliant and insightful critique of the contemporary approach to environmental remediation. Sterling especially gets the way that technology is a *disruptor,* that it unmakes the status quo over and over again, and that a battle of technologies is a battle in which the sands never stop shifting. Casting his tale into the future allows him to illustrate just how uneven our footing is in the present day—and the fact that the book consists of humans getting by, even getting ahead, despite all the chaos and devastation, makes *The Caryatids* one of the most optimistic books I've read in recent days.

Moving back in time, there's William Gibson's *Spook*

Country (Penguin, 2008), a science fiction novel so futuristic that Gibson set it a year *before* it was published. This was a ballsy, genius move, which Gibson characterized as "speculative presentism"—a novel that uses the tricks of science fiction in a contemporary setting, telling a story that revolves around technology and its effect on people. Gibson's protagonist is Hollis Henry, a washed-up pop star who is writing for an art magazine published by a sinister, gigantic PR firm. An assignment brings her into the orbit of a set of post-national spies fighting an obscure and vicious battle, with motivations that are baffling and, eventually, wonderful. Contrasting spy craft, technological art, and the weird, hybrid semi-governmental firm that is characteristic of the twenty-first century, this book makes you feel like you are indeed living in the future, right here in the present.

Go further back to Jo Walton's recently completed Small Change trilogy: *Farthing* (Tor, 2006), *Ha'penny* (Tor, 2007), and *Half a Crown* (Tor, 2008), a series of alternate history novels set in the United Kingdom after a WWII that ended with Britain retreating from the front and ceding Europe to the Third Reich in exchange for an uneasy peace. Now that peace is fracturing, as fascist Europe's totalitarian logic demands that all its neighbors bend to their rules, norms, and laws—otherwise the contrast would make the whole arrangement unbearable. If Europe is persecuting its Jews and allied England is not, then there is an unresolvable cognitive dissonance between the two states, one that can only be resolved by England slipping, bit by bit, into a "soft" totalitarian mirror of Nazi Europe. In this naked parable about the erosion of liberties around the world brought on by America's War on Terror, Walton isn't writing about the past any more than Sterling is writing about the future. Her books are a relentless, maddening, inevitable story of how good people let their goodness dribble away, drop by drop, until they find themselves holding nooses.

Science fiction is a literature that uses the device of futurism to show up the present—a time that is difficult enough to get a handle on. As the pace of technological change accelerates, the job of the science fiction writer becomes not harder, but *easier*—and more necessary. After all, the more confused we are by our contemporary technology, the more opportunities there are to tell stories that lessen that confusion.

A Cosmopolitan Literature for the Cosmopolitan Web

Standing in Melbourne airport on the day before this year's World Science Fiction convention, I found myself playing the familiar road-game known to all who travel to cons: spot the fan. Sometimes, "spot the fan" is pitched as a pejorative, a bit of fun at fannish expense, a sneer about the fannish BMI, B-O, and general hairiness. But there are plenty of people who are heavyset, and practically everyone debarking an international flight to Melbourne is bound to smell a little funky, and beard-wearing is hardly unique to fandom.

If there is one thing that characterizes fandom for me, it is a kind of cosmopolitanism. Now, we tend to think of "cosmopolitan" as a synonym for "posh" or "well-travelled." But that's not what I mean here: for me, to be cosmopolitan is to live your life by the ancient science fictional maxims: "All laws are local" and "No law knows how local it is." That is, the eternal verities of your culture's moment in space and time are as fleeting and ridiculous as last year's witch-burnings, blood-letting, king-worship, and other assorted forms of idolatry and empty ritual.

One of science fiction's greatest tricks is playing "vast, cool intelligence" and peering through a Martian telescope aimed Earthwards and noticing just how weird and irrational we all are. At its best, science fiction is a literature that can use the safe distance of an alien world or a distant future as a buffer-zone in which all mores can be called into question—think, for example, of Theodore Sturgeon's story of the planet of enthusiastic incest-practitioners, "If All Men Were Brothers, Would You Let One Marry Your Sister?" published in *Dangerous Visions* in 1967.

Behind every torturer's mask, behind every terrible crusade, behind every book-burning and war-drum is someone who has forgotten (or never learned) that all laws are local. Forgetting that all laws are local is the ultimate in hubris, and it is the province of yokels and bumpkins who assume that just because *they* do something in a particular way, *all* right-thinking people always have and always will. For a mild contemporary example, consider the TV executive who blithely asserts that her industry is safe, because no matter what happens in the future, the majority of us will want to come home, flop down on the sofa, and turn on the goggle-box—despite the fact that TV has existed for less than a century, a flashing eyeblink in the long history of hominids, most of whom have gotten by just fine without anesthetizing themselves with a sitcom at the end of a long day.

Which is not to say that cosmopolitans don't believe in anything. To be cosmopolitan is to know that all laws are local, and to use that intellectual liberty to decide for yourself what moral code you'll subscribe to. It is the freedom to invent your own ethics from the ground up, knowing that the larger social code you're rejecting is no more or less right than your own—at least from the point of view of a Martian peering through a notional telescope at us piddling Earthlings.

My high-school roommate, Possum Man, was the very

apotheosis of a science fiction cosmopolitan. Educated in the radical (and quite wonderful) Waldorf school system, Possum decided that quantitative grades and credits cheapened the learning process. So even though he took a full roster of courses, he rejected all grades and credits for his (quite excellent) work, and never received a formal diploma despite a long and honorable career in our alternative secondary school.

Possum was willing to reconsider anything and everything from the Martian distance. One day, he noticed that the insides of his knit sweaters were much more interesting than the outsides—busting with tasty asymmetries and pretty loose ends, a topography that was far more complex and chewy than the boringly regular machine-made exterior. From that day forward, he started wearing the sweaters inside out. (Today, he helps coordinate Toronto's free school, AnarchistU.)

Which brings me back to spot-the-fan. Looking for fans isn't just about looking for heavyset people, or guys with big beards, or people who are sloppily dressed. Looking for fans is about looking for people who appear to have given a great deal of thought to how they dress and what they're doing, and who have, in the process of applying all this thought to their daily lives, concluded that they would like to behave differently from the norm. It is about spotting people who are dressed as they are not because of fashion, nor because of aspiration, but because they have decided, quite deliberately, that this is the best thing for them to wear. (Before I go on, let me hasten to add that *some* fans are simply bad dressers with poor hygiene and grooming—but that's hardly the exclusive province of fandom or any other subculture.)

There's something comforting about cosmopolitanism, especially if you start off as someone who's a little bit weird or off-kilter. Cosmopolitanism comforts you with messages like, "The head cheerleader and the quarterback may rule the school, but they have no more virtue than the peacock with the biggest feathers, the goldfish with the bulgiest eyes, and in most of the cultures that ever existed, they would be thought ugly, stupid, and ridiculous." The haughty distance of cosmopolitanism lets you avoid the misery of the daily, earthly reality of being a social pariah—*I may be a Martian, but at least I can look down on all of you from Mars and see your absurdity for what it is.*

And once you start, it's hard to stop. Reading Patterson's recent biography of Robert A. Heinlein, *Learning Curve*, I was struck by how much fringey stuff old RAH dabbled in: telepathy, radical politics, polyamory (or "companionate marriage," as it was called in his day), nudism, and all manner of funny business, all of which is reflected in his books, and all of which can be summed up with "all laws are local."

That takes me to the Web, and to "Rule 34": "If it exists, there is porn of it. No exceptions." (Charlie Stross has recently completed a book called *Rule 34*, which sounds like a hoot). Rule 34 can be thought of as a kind of indictment of the Web as a cesspit of freaks, geeks, and weirdos, but seen through the lens of cosmopolitanism, Rule 34 suggests that the Web has given us all the freedom to consider that the rules we bind ourselves by are merely local quirks, and to take the liberty to turn our sweaters inside out, practice exotic forms of vegetarianism, or have sex while wearing giant anthropomorphic animal costumes.

Rule 34 bespeaks a certain sophistication—a gourmet approach to life. As Kevin Kelly points out in his excellent new book, *What Technology Wants*, a gourmet isn't someone who shovels everything he can get hold of into his gob; rather, it's someone who looks long and hard at all the available options and picks the ones he finds best. Kelly's definition is an important one, because it provides a roadmap to a sophisticated approach to any product or practice; for example, this definition makes the Amish into the world's greatest technophile, since the avant-garde of Amish hackers try every new technology, evaluate whether it fits well into Amish life, and report back to the wider community, who decide whether and how to adopt the tool or service based on what it is likely to do to their lives. While the rest of us are gobbling up new technologies like they were $0.99 Super Big Meals, the Amish are carefully tweezering out the best bits and leaving the rest behind.

Rule 34, the Amish, and fandom's willingness to wear its sweaters inside-out are the common thread running through the 21st century's social transformations: we're finding a life where we reevaluate social norms as we go, tossing out the ones that are empty habit or worse, and enthusiastically adopting the remainder because of what it can do for our lives. That is modern, sophisticated, gourmet cosmopolitanism, and it's ever so much more fun than the old cosmopolitanism's obsession with how they're wearing their cuffs in Paris, or what's on at the Milan opera.

When Love Is Harder to Show Than Hate

Copyright law is set up to protect critics, while leaving fans of creative works out in the cold

When a group of fans of the *Dune* books received a copyright threat from the estate of Frank Herbert, they took the path of least resistance: they renamed and altered their re-creation of the novel's setting—a loving tribute created inside the virtual world of Second Life—so that it was no longer so recognisable as an homage to Herbert's classic science fiction novels.

The normal thing to do here is to rail at the stupidity of the Herbert estate in attacking these fans. After all, they weren't taking money out of the pockets of the estate, the chance of trademark dilution in this case is infinitesimal, the creators were celebrating and spreading their love for the series, they are assuredly all major fans and customers for the products the estate is trying to market, their little Second Life re-creation was obscure and unimportant to all but its users, and the estate's legal resources could surely be better used in finding new ways to make money than in finding new ways to alienate its best customers.

But that's not what this column is about. What I want to ask is, how did we end up with a copyright law that only protects critics, while leaving fans out in the cold?

Some background: copyright's regulatory contours allow for many kinds of use without permission from the copyright holder. For example, if you're writing a critical review of a book, copyright allows you to include quotations from the book for the purpose of criticism. Giving authors the right to choose which critics are allowed to make their points with quotes from the original work is obvious bad policy. It's a thick-skinned author indeed who'd arm his most devastating critics with the whips they need to score him. The courts have historically afforded similar latitude to parodists, on much the same basis: if you're engaged in the parodical mockery of a work, it's a little much to expect that the work's author will give her blessing to your efforts.

The upshot of this is that you're on much more solid ground if you want to quote or otherwise reference a work for the purposes of rubbishing it than if you are doing so to celebrate it. This is one of the most perverse elements of copyright law: the reality that loving something doesn't confer any right to make it a part of your creative life.

The damage here is twofold: first, this privileges creativity that knocks things down over things that build things up. The privilege is real: in the 21st century, we all rely on many intermediaries for the publication of our works, whether it's YouTube, a university web server, or a traditional publisher or film company. When faced with legal threats arising from our work, these entities know that they've got a much stronger case if the work in question is critical than if it is celebratory. In the digital era, our creations have a much better chance of surviving the internet's normal background radiation of legal threats if you leave the adulation out and focus on the criticism. This is a selective force in the internet's media ecology: if you want to start a company that lets users remix TV shows, you'll find it easier to raise capital if the focus is on taking the piss rather than glorifying the programmes.

Second, this perverse system acts as a censor of genuine upwellings of creativity that are worthy in their own right, merely because they are inspired by another work. It's in the nature of beloved works that they become ingrained in our thinking, become part of our creative shorthand, and become part of our visual vocabulary. It's no surprise, then, that audiences are moved to animate the characters that have taken up residence in their heads after reading our books and seeing our movies. The celebrated American science-fiction writer Steven Brust produced a fantastic, full-length novel, *My Own Kind of Freedom*, inspired by the television show *Firefly*. Brust didn't—and probably can't—receive any money for this work, but he wrote it anyway, because, he says, "I couldn't help myself."

Brust circulated his book for free and was lucky enough that Joss Whedon, *Firefly*'s creator, didn't see fit to bring legal action against him.

But if he had been sued, Brust would have been on much stronger grounds if his novel had been a savage parody that undermined everything Whedon had made in *Firefly*. The fact that Brust wrote his book because he loved Whedon's work would have been a mark against him in court.

This isn't a plea for unlimited licence to commercially exploit the creations of others. It's fitting that commercial interests who plan on making new works from yours seek your permission under the appropriate circumstances. Nor is this a plea to eliminate the vital aid to free expression that we find in copyright exceptions that protect criticism.

Rather, it's a vision of copyright that says that fannish celebration—the noncommercial, cultural realm of expression and creativity that has always accompanied commercial art, but only lately attained easy visibility thanks to the internet—should get protection, too. That once an artist has put their works in our head, made them part of our lives, we should be able to live those lives.

Think Like a Dandelion

Regular *Locus* readers will have noted a recent front-of-the-book item about my recent Good News, a little daughter named Poesy, born to us on February 3, 2008. This feat of nanoengineering—mostly accomplished by my Alice, with 23 chromosomes' worth of programming assistance from yours truly—has got me thinking about reproduction, even more than usual.

Mammals invest a lot of energy in keeping track of the disposition of each copy we spawn. It's only natural, of course: we invest so much energy and so many resources in our offspring that it would be a shocking waste if they were to wander away and fall off the balcony or flush themselves down the garbage disposal. We're hard-wired, as mammals, to view this kind of misfortune as a moral tragedy, a massive trauma to our psyches so deep that some of us never recover from it.

It follows naturally that we invest a lot of importance in the individual disposition of every copy of our artistic works as well, wringing our hands over "not for resale" advance review copies that show up on Amazon and tugging our beards at the thought of Google making a scan of our books in order to index them for searchers. And while printing a book doesn't take nearly as much out of us as growing a baby, there's no getting around the fact that every copy printed is money spent, and every copy sold without being accounted for is money taken away from us.

There are other organisms with other reproductive strategies. Take the dandelion: a single dandelion may produce 2,000 seeds per year, indiscriminately firing them off into the sky at the slightest breeze, without any care for where the seeds are heading and whether they'll get an hospitable reception when they touch down.

And indeed, most of those thousands of seeds will likely fall on hard, unyielding pavement, there to lie fallow and unconsummated, a failure in the genetic race to survive and copy.

But the disposition of each—or even most—of the seeds aren't the important thing, from a dandelion's point of view. The important thing is that every spring, *every crack in every pavement is filled with dandelions.* The dandelion doesn't want to nurse a single precious copy of itself in the hopes that it will leave the nest and carefully navigate its way to the optimum growing environment, there to perpetuate the line. The dandelion just wants to be sure that every single opportunity for reproduction is exploited!

Dandelions and artists have a lot in common in the age of the internet. This is, of course, the age of unlimited, zero-marginal-cost copying. If you blow your works into the net like a dandelion clock on the breeze, the net itself will take care of the copying costs. Your fans will paste-bomb your works into their mailing lists, making 60,000 copies so fast and so cheaply that figuring out how much it cost in aggregate to make all those copies would be orders of magnitude more expensive than the copies themselves.

What's more, the winds of the internet will toss your

works to every corner of the globe, seeking out every fertile home that they may have—given enough time and the right work, your stuff could someday find its way over the transom of every reader who would find it good and pleasing. After all, the majority of links between blogs have been made to or from blogs with four or fewer inbound links in total—that means that the internet has figured out a cost-effective means of helping audiences of *three people* discover the writers they should be reading.

So, let's stipulate that you want to reproduce like a dandelion and leave mammaldom behind. How do you do it?

There are two critical success factors for dandelionhood:

1. Your work needs to be *easily* copied, to *anywhere* whence it might find its way into the right hands. That means that the nimble text-file, HTML file, and PDF (the preferred triumvirate of formats) should be distributed without formality—no logins, no email address collections, and with a license that allows your fans to reproduce the work on their own in order to share it with more potential fans. Remember, copying is a *cost-center*—insisting that all copies must be downloaded from your site and only your site is insisting that you—and only you—will bear the cost of making those copies. Sure, having a single, central repository for your works makes it easier to count copies and figure out where they're going, but remember: dandelions don't keep track of their seeds. Once you get past the vanity of knowing exactly how many copies have been made, and

find the zen of knowing that the copying will take care of itself, you'll attain dandelionesque contentment.

2. Once your work gets into the right hands, there needs to be an easy way to consummate the relationship. A friend who runs a small press recently wrote to me to ask if I thought he should release his next book as a Creative Commons free download in advance of the publication, in order to drum up some publicity before the book went on sale.

I explained that I thought this would be a really *bad idea.* Internet users have short attention spans. The moment of consummation—the moment when a reader discovers your book online, starts to read it, and thinks, huh, I should buy a copy of this book—is very brief. That's because "I should buy a copy of this book" is inevitably followed by, "Woah, a youtube of a man putting a lemon in his nose!" and the moment, as they say, is gone.

I know this for a fact. I review a lot of books on Boing Boing, and whenever I do, I link to the Amazon page for the book, using my "affiliate ID" in the URL. If you follow one of those links and buy the book, I get a commission—about eight percent. I can use Amazon's reporting tool to tell exactly how many people click on my links, and how many of them shell out money for the book, and here's the thing: when I link to a book that's out soon, available now for pre-order, I reliably get less than ten percent of the purchases I get when I link to books that are available for sale now. Nine out of ten Boing Boing readers who buy books based on my reviews don't want to pre-order a title and wait for it to show up later.

The net is an unending NOW of moments and distractions and wonderments and puzzlements and rages. Asking someone riding its currents to undertake some kind of complex dance before she can hand you her money is a losing proposition. User-interface designers speak of how every additional click between thought and deed lops a huge number of seeds out of the running for germination.

In my next column, "Macropayments," I'll write more about this consummative act, for this is the key to enduring success as a dandelion. Here's the gist: expend less effort trying to ensure that small sums of money are extracted from your fans for individual copies of your work, and focus instead on getting larger payouts, making each germination count for something more than a buck's royalties.

Digital Licensing: Do It Yourself

Introduction

When someone wants to license your art, characters, photos, articles, or music, how does it shake out? Chances are that these negotiations involve expensive lawyers on both sides of the deal.

If you're running an enlightened company, you might have a Creative Commons license hanging out there for non-commercial, "fannish" uses. (Creative Commons publishes a suite of widely adopted licenses that allow rights-holders to release their work for sharing, remixing, etc.)

But somewhere between Creative Commons and full-blown, lawyerly license negotiation is a rich, untapped source of income for creative people and firms with portfolios of iconic material. To cash in, you just need the courage to let go of a little control.

Read on...

How We Got Here

Before Creative Commons, there were lawyers. If I wanted to make a Mickey Mouse ear-wax scraper (such a thing does exist, and I once lost an eBay auction for it), I'd need to hire a lawyer who was sufficiently high-powered to get Disney's lawyers to return his calls. After several thousand dollars worth of pitching and drafting and arguing, I'd get my license and could go back to my factory and start cranking out my own special brand of cute and hygienic devices. Presuming, of course, that Disney was willing to grant me the license at all.

This approach works reasonably well for certain kinds of products and services. While I'm sure Bayer would prefer not to keep a couple of lawyers on hand to negotiate with Hanna-Barbera every time it wants to change the packaging on Flintstones Vitamins, it's not a great hardship to have them on staff.

Businesses like Bayer know how to talk to businesses like Hanna-Barbera: An electrician might say that they were "impedance-matched"—that is, they speak the same language, employ the same protocols, and have the same base assumptions about how the world should work.

Before the internet, this state of affairs was, broadly speaking, sufficient. If there was such a thing as a "mom-and-pop vitamin manufacturer," they probably wouldn't be so cheeky as to produce their own unauthorized Flintstones Vitamins, and if they did, they'd either be so obscure as to escape notice and commercial success, or they'd rise to the level of corporate notice by Bayer or Hanna-Barbera, who would crush them into paste.

After the internet, it suddenly became possible to be:

- A mom-and-pop, hand-crafted kind of producer; who
- Expected to be able to use trademarks and copyrights; and
- Who rose to the attention of lots of people; but

- Didn't have any money, lawyers, or even business.

This was a genuinely novel situation. Fans whose fan-fiction stories had formerly reached small groups of friends now reached potentially *gigantic* groups of friends. They were visible to search-engines (and hence rights-holders). And, technically, they were liable for enormous statutory damages that had been put in place to deter rival media companies and manufacturers by putting a little sting into their punishment—but that little sting was a devastating blow when applied to individuals.

Enter Creative Commons...

Creative Commons

As a non-profit group that provides several kinds of legal licensing agreements for use by content creators, Creative Commons serves an important purpose for today's Net culture.

Since 2001, Creative Commons has distributed these licenses for free to creators all over the world, undertaking the Herculean task of making the licenses binding in dozens of legal systems.

Commons licenses are clean, standardized, universal licenses that lay out simple rules-of-the-road for "doing culture." They allow rights-holders to clearly communicate a set of permissible uses for their works that are offered to all and sundry.

For example, Creative Commons licenses allow you

to take my novels and copy them, share them, translate them, reformat them, make new works out of them like movies and plays—provided that you do so non-commercially. If you're making money at this, you have to come and get a license.

It works great. My books are published by real, brick-and-mortar publishers who stick real, reasonably priced lumps of paper in real, well-lit stores, where they change hands for real money. My fans, meanwhile, are empowered to do practically any non-commercial thing with the books that they want: Kids make short movies for school assignments; adults translate into foreign languages to hone their language skills; artists do drawings and paintings for the love of it. People podcast 'em, email them to friends, and otherwise have a good time, all the while generating the market for those physical books.

So far, so good.

Questions of Commerce

The internet isn't just full of noncommercial fans and commercial artists, though.

There's a whole continuum of production that the internet has engendered, and quite a lot of it involves money changing hands—something Creative Commons isn't quite equipped to cope with.

Take Etsy, for example, which is among my favorite places in the entire *noosphere*: It's like eBay for crafters. It's filled with innumerable creators who make physical objects and offer them for sale.

What kind of physical objects?

What kind would you like?

Jewelry, clothes, toys, books, sculpture, painting, game controllers, hand-tooled keyboards, masks, furnishings, tableware, collage, drawings, picture frames, musical instruments, tools—every imaginable product of the cunning artificer's workshop. It's like the Olde Curiosity Shoppe, come to virtual life and expanded to infinite size through several spatial and temporal dimensions.

And it's just a corner of the *makerverse.* From edge to edge, the Net is filled with creators of every imaginable tchotchke—and quite a lot of them are for sale.

And quite a lot of *that* is illegal.

That's because culture isn't always non-commercial. All around the physical world, you can find markets where craftspeople turn familiar items from one realm of commerce into handicrafts sold in another realm.

I have a carved wooden Coke bottle from Uganda, a Mickey Mouse kite from Chile, a set of hand-painted KISS matrioshkes from Russia. This, too, is a legitimate form of commerce, and the fact that the villager who carved my Coke bottle was impedance-mismatched with Coke and didn't send a lawyer to Atlanta to get a license before he started carving isn't a problem for him, because Coke can't and won't enforce against carvers in small stalls in marketplaces in war-torn African nations.

If only this were true for crafters on the Net. Though they deploy the same cultural vocabulary as their developing-world counterparts for much the same reason (it's the same reason Warhol used Campbell's soup cans), they don't have obscurity on their side. They live by the double-edged sword of the search-engine: The same tool that enables their customers to find them also enables rights-holders to discover them and shut them down.

It doesn't have to be this way.

The Alternative

Creative Commons works because all you have to do to "license" a work for re-use is to follow and link and read three or four bullet points.

It is *impedance-matched* with Net culture.

Lawyer-licensing doesn't work for makers, because hiring a lawyer to discover if you can net $45 selling three $28 t-shirts is not cost effective. Even assuming you can get the license, you'd have to raise the cost of the t-shirt to $450 to cover the lawyer-time incurred in getting it.

What would an impedance-matched licensing regime look like for makers? A lot like Creative Commons.

Here's one model: Imagine if you included the following text alongside all your logos, literature, photos, and artwork:

"You are free to use the visual, textual, and audiovisual elements of this work in commercial projects, provided that you remit 20 percent of the gross income arising from your sales to doctorow@paypal.com. You are required to remit these funds on a quarterly basis, or on an annual basis where the total owing is less than $100."

That's it. For extra effect, put it on a Webpage with downloadable high-resolution artwork, source videos, 3D meshes—whatever the preferred form of a work for modification might be.

Oh, you could hire a lawyer to tart it up a little. There's probably a business in this for someone who wants to found a firm devoted to fine-tuning the language to ensure it works in multiple jurisdictions and who wants to act as a payment clearing house.

But the point of this license is that it is primarily *normative*—that is, it's a discussion between two civilians (you and some potential crafter) about some reasonable rules of the road.

Complexity is your enemy here. Two or three sentences are all you want, so that the idea can be absorbed in 10 seconds by a maker at three in the morning just as she embarks on an inspired quest to sculpt a 3D version from your logo using flattened pop-cans.

The secret to simplicity here is in the license fee, the payment schedule, and the enforcement regime.

The Self-Serve Difference

A lawyerly license usually generates a fairly small per-unit royalty on a lot of sales—say, 5 to 10 percent—and is front-loaded with an initial payment. By charging a *much* higher per-unit royalty and waiving the upfront fee, a maker can take your license on with almost no risk.

He can, for instance, sculpt a steampunk assemblage of your mascot and simply raise the price on the final item a little to cover part of your cut. He still carries the usual risk associated with making art without having a buyer lined up in advance, but he doesn't have to worry that after finishing it, you'll come along and threaten him with a lawsuit.

How do you enforce this license? You don't.

Or rather, you do, but only when it's worth it. Chances are you're not enforcing against most of the little guys these days, because you haven't heard of most of them—and when your lawyers send threatening letters to beloved Etsy sellers who want to celebrate your products, it makes you look like a goon.

If you're sane and smart, you save your enforcement efforts for the Big Guys, people who are clearly living beyond the hand-to-mouth existence of a cottage crafter, firms that list a bunch of regional distributors, and so forth.

Self-serve license enforcement works exactly the same way. You assume that most people are honest and want to do the right thing (a surprising number of people are, especially when the right thing is easy and impedance-matched). When you find little penny-ante chiselers making out like bandits, ignore them. There are only so many hours in the day, and you're better off spending them ensuring that everyone who wants to pay you can, rather than wasting your time ensuring that everyone who uses your stuff pays.

When you find the big operators, you pay lawyers to threaten them, just as you do now.

The only difference is that honest people have a way to pay you that makes sense for you, and sense for them.

There's one other difference between lawyerly and self-serve licensing: By allowing a much wider diversity of authorized products to exist than could possibly flourish under a top-down, command-and-control regime, you get a free way to discover the opportunities that never occurred to you.

Every crafter becomes your researcher, bearing all the costs of market-testing every conceivable variation on your product. When something starts to really sell, you can bring the crafter in-house by bringing out the lawyers and negotiating a cheaper license for her that gives you more direct control over the production and quality.

A Built-In Future

But what about the brand, the trademark, the almighty *image?*

The brand is easy. Add a condition to your license:

"As a condition of this license, your work must prominently bear the SELF-SERVE LICENSING logo and the words: THIS WORK IS CREATED UNDER THE TERMS OF A SELF-SERVE CRAFTER'S LICENSE. THE ORIGINAL CREATORS FROM WHICH THIS IS DERIVED HAVE NOT REVIEWED IT OR APPROVED IT, THOUGH THEY ARE COMPENSATED FOR ITS SALE."

That's the whole proposal: two paragraphs of simple, plain-language text and a little, easily recognizable logo, and you'd get yourself a whole world of cheap and easy licensing that would turn yesterday's pirates into tomorrow's partners.

This has the neat effect of satisfying the trademark question, too: Notices like this preserve the integrity of the trademark, ensuring that customers are continuously notified about the relationship of your marks and your authorization, protecting you from legal dilution.

It's inevitable that some junk will emerge from this stuff, some of which will embarrass you. But Creative Commons showed that cultural and commercial culture could exist alongside one another; a self-serve licensing system aimed at bringing Creative Commons to commercial transactions shows that artisans and commerce can enjoy the same mutually beneficial relationship.

New York, Meet Silicon Valley

There's no progress to report on the short story collection this month, beyond a key lesson about being the writer and publisher of a book—when "the writer" goes on a book tour, "the publisher" doesn't get to do production work.

As I noted in last month's column, early time-line slips in the production process for *With a Little Help* pushed me into the time that I'd earmarked for a tour for my "real" publishers (Tor Teen and Harper U.K.) supporting my most recent novel, *For the Win*. Of course, that lesson is just the kind of thing I set out to learn when I started this project: which parts of the process can be handled by moderately business-savvy, productive writers, and which parts need publishers, packagers, managers, and other helpers for.

The broader premise of my experiment, of course, is that the internet changes things. Specifically, the internet makes it cheaper to coordinate complex tasks than ever before. This is the revolutionary thing about information technology: it can automate coordination, enabling things that used to be expensive and complex to be cheap and simple. The other thing technology makes cheap is experimentation. That's the special genius of IT-based projects: you can fail cheap.

Look at it this way: starting a magazine is hard. It costs money. Magazine founders mortgage their houses, convince their friends to quit their jobs and move across the country, print letterhead, fell trees, pulp them, and cover them with toxic, heavy metal–based inks. It's the kind of thing you want to be really sure about before you take it on. If 75% of the people who attempted to start a magazine abandoned the venture in a few weeks, it would represent a tremendous waste of time and money.

By contrast, most people who start blogger or Twitter accounts may very well abandon them. But starting up a blogger or Twitter account takes about five minutes. And they cost nothing. It's the kind of thing you can experiment with in your spare evenings, after the kids are in bed, and the kind you can fail at without losing anything.

Too Cheap to Fail

It's a good thing that IT makes failure so cheap, because IT also creates a range of possible futures so large that it's almost impossible to guess right about which direction to try first. Successful IT innovation is almost never a matter of accurately predicting the future and then building the business that future demands. Successful IT innovation looks more like this:

Once upon a time, two nice folks started a company that made Game Neverending, a whimsical, multiplayer Flash game. They got a bunch of interesting people on their advisory board and on their alpha and beta-test teams. A small squad of dedicated, in-house programmers avidly watched what their players did, around the clock, and they changed the game often, sometimes as often as every 30 minutes, adding, removing, or tweaking features, and watching the players' reactions.

One wildly popular feature right off the bat was an image-sharing system that let players show their friends pictures they took or found. Here's my claim to fame: I asked for this feature, because the woman I was dating lived in London, and I lived 9,000 miles away in San Francisco, and it was too cumbersome to share pictures from our days by email.

As the Game Neverending team kept tweaking that image-sharing feature, the players went nuts with it, creating a feedback loop that eventually led to the image-sharing feature taking over the game entirely. Game Neverending ended. But the product lived on, and it was renamed Flickr. (And, by the way, I'm now married to the woman, we have a daughter, and we live in London.)

Writers know how this process works, too. You start with an idea for a book. You roll it around in your mind. You "beta-test" it on your friends, pitch it to your editor and agent. And every time you describe it, it changes a little based on what you learned the last time you talked about it. It costs nothing to change the way you describe your nonexistent book. And, iteration by iteration, your kernel of an idea germinates into something that you'd never have predicted when you sat first sat down to write.

On the other hand, no sane writer would dream of recasting her book as a completely different project after it's been turned in, gone off for copyediting, and been put into production. That's when experimentation goes from cheap to expensive. That's when you start learning the hard way.

This marks a key difference between New York publishing and Silicon Valley. Unlike New York publishing, Silicon Valley's products remain experimental long after they reach the marketplace. Google can change its search layout in seconds flat, try it out on a million searchers, crunch the data, revise the experiment, and do it again, a hundred times a day if they wish. And bad ideas can be just as interesting as good ideas, because when it doesn't cost anything to find out how bad an idea is, you can afford to be pleasantly and enormously surprised when it turns out that, say, people really do want to play Pac-Man on their search-results page.

I consider *With a Little Help* to be a Silicon Valley experiment. My upfront costs are minimal. I've spent $2,256 getting into production, and taken in about $14,400 in payments. I'll probably spend another $200–$300 before I ship, and that's the last money I should have to spend without taking in money first: every time someone buys an on-demand book from Lulu, I'll get paid without expending any capital. I'm printing and binding my short-run hardcovers in lots of 20, after being paid for them. The audiobook CDs are also produced on-demand by a third party, which means no capital costs for me, either. Setting up the donation page took a few hours fiddling with PayPal, and even if I never take in a penny in donations, I'm not out a penny either.

The "Standard" Response

This is what Silicon Valley can teach New York: make experiments cheap. Don't hire a pricy, outsourced IT company to design a new, exciting book-service for your company. Why not hire your own developers and a visionary tech person and try something? Wait until it fails, learn from that failure, and try something else. Your outsource IT company will hate you if you call them every 30 minutes asking to try a new feature, or to tweak or remove an existing one. But your in-house people will love the challenge and freedom of being allowed to fail fast, iterate, and learn.

Here's what Silicon Valley can't teach New York publishers: how to prevent copying. Last month at BEA, publishing CEOs all but begged Silicon Valley to present them with a universal, interoperable DRM system that would prevent copying without locking books to one vendor's platform. "Our fondest wish is that all the devices become agnostic so that there aren't proprietary formats and you can read wherever you want to read," Penguin's David Shanks reportedly said.

If you ask a few big tech companies to "standardize" a format for your e-books that others can only implement with their permission, they'll happily start planning how to spend the money they're about to make off you. But DRM is incompatible with the idea of standardization—that's why Silicon Valley loves it. Because lurking in the heart of every entrepreneur is a monopolist hoping to shut out the competition.

On the other hand, formats don't matter when there's no DRM in the mix. Take for example the *Publishers Weekly* homepage. As of this moment, it contains embedded objects in six different formats, ranging from JPEG to HTML. As a reader, I don't have to know, or care, whether the PW logo is a GIF, a PNG, or a BMP because there are practically no restrictions on renderers for any of these formats. Any programmer who wants to make a browser can go to a consortium's Web site, grab some reference code for displaying its format, and massage it into her software. She can tweak the code, refactor it completely, or just pay attention to the parts that she cares about.

That's how standards work. Just like standard-gauge rails opened the continent to trains because they never specified whose engines could run on them, or what kind of freight they could pull, universal standards for e-books developed by publishers could do the same for the reading landscape of e-book readers, tablets, and e-books.

Universal standards from real standardization bodies like ISO (International Organization for Standardization), the W3C (World Wide Web Consortium), the IETF (The Internet Engineering Task Force), or the IDPF (International Digital Publishing Forum) would still attract all the tech giants, but they would also attract everyone else, from zippy, ADD-addled startups to copyright holders and activists—everyone with a stake in the outcome. These organizations will make you a standard, like epub, for example. It might not be adopted the first time around. But that's OK. Because you'll make another, and then another. And without DRM, readers who bought your books in the first formats wouldn't have to worry if a standard dies and is replaced by another.

There is a lesson for publishers in how giants like Silicon Graphics, AltaVista, and Commodore were beaten into the dirt by snot-nosed startups that used the low cost of experimentation to outcompete them. Publishers should take a page from those upstarts' playbooks. The cool thing about Silicon Valley's brand of experimentation is that failure is often just slow success. And as every good entrepreneur knows, the best way to double your success rate is to triple your failure rate.

With a Little Help: The Price Is Right

First, a progress report: *With a Little Help* is going great guns behind the scenes. The typographer has some very nice samples for me, and the book should be in my hands—or on my hard drive—shortly. The sound editor's nearly done with the audiobook, which has a lovely handcrafted quality, thanks to all the various environments in which it was recorded. And I'm dutifully uploading gigabytes of scanned paper to Flickr as part of the special edition.

Meanwhile, the mysteries of price and profit are on everyone's minds these days thanks to the Macmillan-Amazon spat, with commentators on both sides of the debate drawing parallels to the train wreck of a decade the recording industry just went through. Those rooting for Macmillan point to the way listeners allegedly abandoned their willingness to pay for music—even as a single retailer, Apple, gained near-total control over pricing and distribution. Those who take Amazon's side point to the recording industry's unwillingness to partner with innovative technology firms like Napster, which offered the RIAA a blank check in exchange for a license to continue operating. They also point to Apple's simplified, 99 cent/track pricing as the breakthrough that listeners needed to start paying.

I think they're both right. On one hand, Macmillan should be worried about losing control of its destiny, as Amazon, a single distributor, seeks to lock readers into its devices and services. But on the other hand, Amazon's optimistic (or, some would say, cutthroat) pricing on the cream of the publishing industry's profits—frontlist hardcovers—isn't necessarily a loser for publishers, and it's possible that the world's largest online bookstore just might have some insight into purchasing patterns that publishers need to hear.

What's Your Theory?

Amazon's $9.99 Kindle price, in part, represents a wager that there are enough new readers for frontlist hardcover books that Amazon (and the publishers whose wares it sells) will make up the lost profits from lower prices with greater sales volume. Macmillan's concern is due, in part, to the indisputable fact that the people who shell out good money for an e-book reader are often precisely the kind of price-insensitive consumers upon whom publishing relies to buy books at full price. It all comes down to which profit-maximizing strategy you favor: price discrimination or demand elasticity.

In my last column, I discussed price discrimination: the idea that you make more money by segmenting your customers based on how much they're willing to spend. At the extreme end of price discrimination, you have the airlines, whose opaque pricing is the bane of travelers who can't figure out why a ticket that departs a day earlier costs twice as much. In publishing, price discrimination is accomplished through "windowing." Traditionally, the hardcover comes out first, at the highest price, so price-insensitive customers, whose thrift is outstripped by their impatience, are enticed to shell out. Once that market is exhausted, the paperback comes along, and price-sensitive customers put their money in the pot. Some customers, of course, would buy the hardcover regardless of whether there was a cheaper option available, but publishers (rightly) believe that if paperbacks and hardcovers went on sale on the same day a sizable fraction of the hardcover market would buy the cheaper paperback. Thus, if low-cost e-books are released simultaneous with the hardcover, there's reason to worry that Kindle and iPad owners (big spenders who might otherwise buy the premium item) will prefer to download cheap, convenient e-books.

Demand elasticity is the straightforward idea that new customers will come into your shop if you lower prices. The publishing industry already practices some demand elasticity: new hardcovers, for example, are priced at $27, not $75, because the higher margin at $75 would not make up for the lost sales from readers unwilling to pay the higher price. Many internet companies made their fortunes on demand elasticity. Google, for example, bet that charging less for ads (and using clever automation to make money even on extremely cheap ads) would attract so many new advertisers that they would realize a substantial profit.

Everyone with a product to sell practices both price discrimination and demand elasticity in varying degrees. But when the product you're selling is digital, the correct ratio of one to the other becomes a lot harder to calculate. If you're selling hard goods, whether books, shovels, or coffee beans, the math is easy: you can't make money if you drop your price below the marginal cost of production. But digital goods, like e-books, have almost no marginal costs. Things like credit card processing fees, electricity and bandwidth, and a few other considerations keep the cost from truly falling to $0, but the low marginal cost of selling digital copies opens up some very exciting possibilities for publishers. Could the pool of people willing to buy books—the total number of regular readers—be increased by dropping the price? And could that increase in new customers be large enough to offset losses from smaller margins? Amazon clearly thinks so.

Market Theory

But pricing and profit-maximizing strategies aren't the whole story. Consumer electronics buyer demographics tilt heavily to the coveted 18–34-year-old who'll buy anything slim with an eggshell finish. Turning those big spenders into readers is an exciting prospect for anyone who cares about bringing in new business—and Macmillan executives are keenly aware of the opportunity e-books represent for turning nonreaders into new customers. Tom Doherty, publisher of Macmillan's Tor imprint (Tor publishes my novels), is positively luminous on the importance of inducting nonreaders into the practice of regular reading. And there's no bookseller on earth with more nonreader customers than Amazon, which, in addition to books sells everything from server space to freeze-dried steaks, sex toys, and uranium ore.

Yes, the publishing industry needs to attract new readers. But as the recent skirmish over price suggests, the question is: at what cost? At the heart of the Macmillan-Amazon spat is the realization that allowing Amazon to dominate the e-book market will only make it harder for publishers to balance their interests with Amazon's. That's because the Kindle is a "roach motel" device: its license terms and DRM ensure that books can check in, but they can't check out. Readers are contractually prohibited from moving their books to competing devices; DRM makes that technically challenging; and competitors are legally enjoined from offering tools that would allow readers to break Kindle's DRM and move their books to other devices. Price conflict aside, this is the real challenge for publishers, because it means that e-book customers can't break with Amazon without jettisoning their digital libraries.

Amazon refused to allow any changes to its terms for my last book, both in the Audible edition and the Kindle edition, refusing to allow me to offer the book with some introductory text affirming readers' rights to move the books to devices that Amazon hasn't approved.

Don't hope for a better shake from Apple, either. Apple's longstanding love-affair with proprietary formats and lock-ins will very likely make the iPad every inch the roach motel that the Kindle is. Apple pitches this as a design decision, but it's also a powerful anticompetitive strategy that raises the cost of switching to a competitor's device.

There are other forms of market dominance, too. Amazon has the internet's best affiliate program. Bloggers, or anyone with a Web page, really, can get an affiliate ID from Amazon and use it in their links to Amazon's products. Amazon pays a commission for everything that a customer you send its way buys. For example, a customer who follows a link to a book and goes on to buy a television earns you a tidy sum that Amazon pays out once a month. I regularly review books and products on Boing Boing and use my personal affiliate ID to link to Amazon. In 2009, I sold more than 25,000 books that way, at a commission to me of 4% to 8.5%. It's not the 40% discount I'd get if I was buying books wholesale from Macmillan and selling them in a bricks-and-mortar store, but I don't have any overhead, bookkeeping, cash register, employees, or other expenses.

There's a reason that the Web is festooned with links to Amazon: it pays to make those links, and it's easy. Other retailers, including Indienet, Powell's, Borders, Barnes & Noble, and the amazing Book Depository have their own affiliate programs, and I'd happily link to those, too, if there was an easy way of doing so without having to laboriously hand-code six links on every review. This is every bit as important as DRM, Kindle pricing, and restrictive license terms. Price may be the hot issue now, but publishers should be thinking about the whole picture. An automated system for offering readers more choice in their book buying would also help correct the current imbalance in the e-book market, while improving the lives of book buyers and those who make links on the Web. WordPress's Booklinker plug-in is a good start on this—you can install it on your server and it turns all your book links into a pop-up with various retailers that readers can choose from, and your affiliate ID is automatically added to each URL. Publishers who want to get a jump on Amazon could choose to expand Booklinker by turning it into a Java-Script library that bloggers can include on their Web pages without having to install server software and can use with systems other than WordPress. For extra points, they could figure out how to tie the service into the ISBN resolving services used by libraries to automatically find other editions of a book as well.

Amazon has done an incredible job of figuring out how to cross-sell, upsell, and just plain sell books. They have revolutionized bookselling over the course of a decade. As a reader and a writer, and as a publisher and a bookseller, I am constantly amazed at how good they are at this. But I don't believe in benevolent dictators. I wouldn't endorse a lock-in program run by a cartel of Santa Claus, the Tooth Fairy, and Mohandas Gandhi. As good as Amazon is at what it does, it doesn't deserve to lock in the reading public. No one does.

You Shouldn't Have to Sell Your Soul Just to Download Some Music

The activities that are restricted by download licence agreements range from the ridiculous to the dubious

Here's the world's shortest, fairest, and simplest licence agreement: "Don't violate copyright law." If I had my way, every digital download from the music in the iTunes and Amazon MP3 store, to the ebooks for the Kindle and Sony Reader, to the games for your Xbox, would bear this—and only this—as its licence agreement.

"Don't violate copyright law" has a lot going for it, but the best thing about it is what it signals to the purchaser, namely: "You are *not* about to get screwed."

The copyright wars have produced some odd and funny outcomes, but I think the oddest was when the record industry began to campaign for more copyright education on the grounds that young people were growing up without the moral sensibility that they need to become functional members of society.

The same companies that spent decades telling lawmakers that they were explicitly *not* the guardians of the morality of the young—that they couldn't be held accountable for sex, drugs and rock'n'roll, for gangsta rap, for drug-fuelled dance-parties—did a complete reversal and began to beat their chests about the corrupting influence of downloading on the poor kiddies.

Well, they got it half-right: the fact that kids—and lots of adults—don't see anything wrong with destroying the record labels is certainly bad news for the record companies. Back when Napster started, the general feeling was that the record companies deserved to die for all the packaged boy bands, for discontinuing the single, for killing the backlist, for price-fixing CDs, and for notoriously miserable contracts for artists.

Then came the digital rights management, the lawsuits (first against toolmakers like Napster, then against tens of thousands of music fans), then the use of malicious software to fight copying, the procurement of one-sided laws, the destruction of internet radio. Brick by brick, the record companies built the moral case for ripping them off (and the movie companies, broadcasters, e-book publishers, and game companies weren't far behind). As the copyfight wore on, wrecking the entertainment industry became an ever-more attractive proposition.

A decade later and the record industry has finally brought back the single, and there seems to be some semblance of price-competition (contracts for artists and the existence of boy bands still go in the minus column of course). They've even got rid of digital rights management for the majority of music sales, and the backlist is much bigger than it was in the record-store days.

So now the pitch goes: "We gave you what you asked for, you've brought us to our knees. Now, please stop ripping us off and start buying music again—we're offering a fair deal." But anyone who examines the pitch closely can see it for what it really is: just more bait for yet another trap.

It's that pesky user-agreement. When you go into

one of the few remaining record stores, there's no clerk beside the till chanting, "By buying this music, you agree to the following terms and conditions," rattling off an inexhaustible set of rights that you're surrendering for having the temerity to buy your music instead of ripping it off.

If the sales-pitch for a download is "a fair deal," then it has to *be* a fair deal. The activities that these licence agreements restrict range from the ridiculous to the dubious, though I suppose reasonable people might disagree about the fairness of selling or loaning out your digital music collection.

But it's not the entertainment industry's job to tell me what are and are not fair terms of sale for my downloads. If loaning an MP3 should be illegal, let them get a law passed (they're apparently good at that—the fact that they haven't managed it to date should tell you something about the reasonableness of the proposition). The one-sided, un-negotiated licence agreement lurking behind the "Check here to affirm that you have read and agreed to our terms of service" represents a wishful (even delusional) version of how a purchase works.

If the pitch is, "this is a fair deal," then the EULA should be: "You can do anything with this, so long as you don't break the law."

Not "Abandon hope, all ye who purchase here."

Net Neutrality for Writers: It's All About the Leverage

Imagine this: you pick up the phone and call Vito's, the *excellent* pizza joint down the road where your family's gotten its favorite pepperoni and mushroom every Friday night for years. The phone rings once, twice, then:

"AT&T: The number you have called is not engaged, but the recipient has not paid for premium service. Please hold for 30 seconds, or press 'one' to be connected to Domino's immediately."

This is not an analogy to the Net Neutrality fight. This is an analogy to the *"compromise"* most governments and regulators (including the U.S. Federal Communications Commission) are planning for the internet. In their view, internet service providers should be allowed to "manage" and "traffic shape" their networks to slow down packets from the sites you're connecting to, *provided they disclose that they are doing it*. In the view of the world's regulators, this is the *best* we can hope for from our telecomm policy in the 21st century.

The carriers, of course, hate this. They call it nanny-state regulation. In their view, telecomm companies should be free to retard the packets you request in perfect secrecy, as part of a larger strategy to blackmail websites and web-services into paying bribes for the privilege of access to "their users" (that is, you and me).

This is pretty crummy news from the point of view of

J. Random internet user, but it's even worse for writers and other creators.

How do successful writers use copyright? As negotiating leverage. Once you're a successful, non-commodity writer—that is, a writer whose mere name can sell books and whose work can't be freely interchanged in the publisher's catalog or on the bookseller's shelf with another writer's work—copyright becomes a moderately useful tool for extracting funds from publishers. Copyright becomes a productive club-with-a-nail-through-it with which to threaten publishers who might consider publishing a well-known writer's work without her permission. Likewise, copyright is a useful tool for publishers to use in threatening each other, should one publisher take it into his head to copy a competitor's copyrighted books and sell them. Because of this, a successful writer can even auction her copyrights off between more than one publisher.

But just because copyright can be used for leverage some of the time, by some people, it doesn't follow that it will always provide leverage: for example, you could give unknown writers hundreds of years' worth of copyright, and it wouldn't extract one more penny from any publisher, anywhere in the world. Think of poets: you could give every poet in the world a personal poet's disemboweling pike of copyright enforcement, and it wouldn't raise the word-rate for poetry. Copyright is only useful when it provides leverage; the rest of the time, it's a creator's vestigial appendix (at best) or a nagging hindrance (at worst).

Creators need leverage, and policies, technological changes, and laws that create leverage for artists result in more artists making more money. Contrariwise, changes to the law or technology that take away creators' leverage end up doing real harm to creators' fortunes.

An open, neutral internet is one where anyone can start a kick-ass publishing platform merely by coming up with a good idea. Tim Berners-Lee famously invented the Web from his desk at CERN in Geneva as a tool for sharing scientific papers. Merely by distributing web-browsers and web servers, TBL was able to invent his revolutionary publishing platform. Notably, he didn't have to deploy an army of corporate negotiators to book meetings with suits at telecoms around the world and work out under what terms every ISP would (or would not) allow the WWW to traverse its lines. Unsurprisingly, Berners-Lee is a staunch advocate of Net Neutrality.

Likewise, the creators of YouTube were able to simply kick-start the biggest, most successful video watching—and distributing—platform the world has ever seen merely by *inventing it* and shoving it out the door. They built it, we came, and no phone company got a veto over our desire to watch YouTube.

This delirious world of fast, unfettered invention has delivered untold leverage to creators. Publishers—and studios and record labels—used to be the only effective way to reach a large audience, the only way to extract money from them, the only way to distribute creative works to them. As a result, only a few very lucky, very resourceful creators were able to forgo the entertainment giants and strike out on their own. The rest of us had to take whatever they'd offer and like it (at least until we got big enough to make them bid against each other).

You don't need to self-publish to get a better deal from a publisher or other gatekeeper: you merely need to be *able* to self-publish. A negotiation in which the two choices are "Do it my way" and "Go pound sand" is not one that will end well for the supplicant. The mere existence of a better option than "Go pound sand" raises the floor on our negotiations.

In other words: because the internet had opened up the possibility of a myriad of companies, individuals, and co-ops providing distribution, audience, and income to artists, the old, established institutions now have to compete with someone other than each other, at least at the bottom of the market. And since most artists spend most of their careers at the bottom of the market, the largest benefit you can deliver to the arts is to create a whole chaotic marketplace of services and platforms clamoring for their works.

Not that the telecoms really care about this. Art, schmart. They just want to get paid, and paid, and paid. First they get paid when a company like Google buys a heptillion dollars' worth of internet access for a service like YouTube. Then they get your $10–$80/month for your home broadband. Then they get paid a third time by charging Google to send bits to your broadband link.

But the entertainment giants aren't all that upset by the idea of having to pay twice to access their audience. For one thing, they can afford it. That's what the "giant" in "entertainment giant" means. But more importantly, that's how they've *always done it.* Fanning out a horde of business-development glad-handers to sort out the details of distribution deals with disparate channel operators around the world is second nature for them. There's a floor of their corporate headquarters devoted to this kind of thing. They've got their own annual picnic and everything.

Two-gals-in-a-garage do not have this asset. They have two gals. They have a garage. If Net Neutrality is clobbered the way the telecoms hope it will be, the next Web or YouTube won't come from disruptive inventors in a garage; it will come from the corporate labs at one of the five big media consortia or one of a handful of phone and cable companies. It will be sold as a "premium" service, and it won't upset anyone's multimillion-dollar status quo.

More immediately: if the only way to use the internet to widely and efficiently distribute creative work is to convince a big media company to carry it on its "premium" service, kiss your artistic negotiating leverage goodbye. While artists have been going bonkers over threats to copyright, the media titans and the telecomm ogres have quietly formed a pact that will establish them as permanent gatekeepers to the world's audiences.

Not because reaching those audiences is difficult or technically challenging, but because they've sewn up the market.

And hey, Google must have finally grown up, because they just filed a joint brief with Verizon to the FCC saying non-Neutral networks are OK with them—why not? It's not as if Google will have trouble paying the danegeld. And the next Google will have to raise the capital to bribe the world's ISPs before they can even set up shop.

Meanwhile: every telecomm company is as big a corporate welfare bum as you could ask for. Try to imagine what it would cost at market rates to go around to every house in every town in every country and pay for the right to block traffic and dig up roads and erect poles and string wires and pierce every home with cabling. The regulatory fiat that allows these companies to get their networks up and running is worth hundreds of billions, if not trillions, of dollars.

If phone companies want to operate in the "free market," then let them: the FCC could give them 60 days to get all their rotten copper out of our dirt, or we'll buy it from them at the going scrappage rates. Then, let's hold an auction for the right to be the next big telecomm company, on one condition: in exchange for using the public's rights-of-way, you have to agree to connect us to the people we want to talk to, and vice-versa, as quickly and efficiently as you can.

Here's something every creator, every free speech advocate, every copyright maximalist, and every copyfighter should agree on: allowing the channels to audiences to be cornered by a handful of incumbents is bad news for all of us. It doesn't matter that the lame-duck, sellout FCC won't stand up for us. It doesn't matter that Canada's CRTC and the UK's Ofcom are no better, that regulators around the world are as toothless as newborns. This is the big fight for us—the fight over who gets to decide who will be heard and how.

Proprietary Interest

Last week, I found myself wide awake in bed next to my wife, mulling over an email I'd gotten just before lying down (checking email before bed being as bad a habit as eating before bed—both of which I'm trying to stop doing).

The email came from a very nice person who co-curates one of my favorite internet resources: a LiveJournal group devoted to scanning, posting, and discussing old advertisements, mostly print ads, though there's the occasional YouTube embed showing an old TV or radio ad. Ads are funny things, a real window into the zeitgeist. SF writers have long understood the trick of placing some well-chosen ads in scene as a way of showing the reader what kind of world she's following us into (this goes double for SF film-makers).

But for the longest time, adverts were considered unworthy of preservation. In the pre-1976 era in which U.S. copyright law required formal registration for copyright, it was unheard-of for companies to register advertisements and deposit archival copies with the Library of Congress, which housed all registered works. In many cases, advertisements were omitted from microfilm/microfiche preservation in order to save money. But any time I've happened upon some paleo-magazine, I inevitably find the ads far more interesting, timeless, and provocative than the articles they ran alongside of.

So these amateur archivists are doing wonderful work, spelunking in mountains of thrifted and hoarded print sources for the weirdest, funniest, most charming ads of yesteryear and sharing them with each other and the world. I'm routinely moved to copy these ads to the blog I co-edit, Boing Boing, often with some snappy commentary, and always with a link back to the source, which is considered foundational good manners in blogdom.

Which is how I came to be mulling over an email in bed. It came from one of the group's moderators, and she wasn't pleased with me. She'd gotten the mistaken impression that I had been putting the ads on Boing Boing without crediting their source, and wanted me to improve my behavior. She asked me how I'd feel if I someone took my work without credit, and suggested that I might even consider asking permission from the original posters before I took their scans for my own.

(Before I go further, let me state for the record that this was all a minor misunderstanding that was quickly and amicably and reasonably resolved, and that I have nothing but good will and good wishes for the Vintage Ads group and its hardworking moderators and participants.)

This note had me thoroughly bemused and somewhat befuddled. Manners are all well and good, but the note seemed to miss an important fact: it was advocating a standard that, if applied evenly, would lead to the extinction of the group itself.

By and large, the ads in this marvelous community are in the public domain. This means that they are not copyrighted, cannot be copyrighted, and that no one has more claim on them than anyone else. The public domain is all around you: all the words in our language, all the works published by the U.S. government, all of Shakespeare, all of Dickens, all of Wells, Verne, Austen, etc. It's our collective inheritance, the limitless resource from which all may draw: Disney can use it to make *Snow White and the Seven Dwarfs*, and so can I, and so can you. Sometimes, we do good things with the public domain (being married to an Alice, I have a passion for wonderful *Alice in Wonderland* editions, and there are many of these). Sometimes, we do stupid things with it (Mr. Burton, I'm looking at your *Alice* adaptation in particular). But no matter what we do with it, it endures, and all and any may use it as they wish.

Scanning a public domain item does not attract a new copyright to it. Copyright rewards creativity, not "sweat of the brow." Of course, it's only natural to feel a proprietary instinct to the product of one's labor, but in this case, it's misplaced—or at least, best kept to oneself.

Any ethical claim to ownership over a scan of a public domain work should be treated with utmost suspicion, not least because of all the people with stronger claims than the scanner! To be consistent with the ethical principle that one should never use another's work without permission (regardless of the law or the public domain), every scanner would have a duty to ask, at the very least, the corporations whose products are advertised in these old chestnuts (the very best of them are for brands that persist to today, since these vividly illustrate the way that our world has changed—for example, see the very frank Lysol douche ad). For if scanning a work confers an ownership interest, then surely *paying for* the ad's production offers an even more compelling claim!

And the publishers of the magazines and the newspapers—to scan is one thing, but what about the firm that paid to physically print the edition that we make the scan from? And then there are the copywriters and illustrators and their heirs—if scanning an ad confers a proprietary interest, then surely creating the ad should give rise to an even greater claim?

We do acknowledge these claims, at least a little. A good archivist notes the source. A good critic notes the creator. But that is the extent of the claim's legitimacy. If we afford descendants and publishers and printers and commissioners their own little pocket of customary right-of-refusal over their works, we would eliminate the ability to keep these works alive in our culture. For these owed courtesies multiply geometrically—think of the challenge of getting all of Dickens' or Twain's far-flung heirs to grant permission to do anything with their ancestors' works. What a lopsided world it would be if ten seconds' scanner work with the public domain demanded 100 hours' correspondence and permission-begging to be "polite!"

The right to reproduce the public domain is a bargain: you get to make your copies for free, and owe no one anything. But you also get no claim over your reproductions. To assert otherwise is a suicide-pact, for no practice as purely great as the Vintage Ads community would survive such a principle.

"Intellectual Property" Is a Silly Euphemism

"Intellectual property" is one of those ideologically loaded terms that can cause an argument just by being uttered. The term wasn't in widespread use until the 1960s, when it was adopted by the World Intellectual Property Organization, a trade body that later attained exalted status as a UN agency.

WIPO's case for using the term is easy to understand: people who've "had their property stolen" are a lot more sympathetic in the public imagination than "industrial entities who've had the contours of their regulatory monopolies violated," the latter being the more common way of talking about infringement until the ascendancy of "intellectual property" as a term of art.

Does it matter what we call it? Property, after all, is a useful, well-understood concept in law and custom, the kind of thing that a punter can get his head around without too much thinking.

That's entirely true—and it's exactly why the phrase "intellectual property" is, at root, a dangerous euphemism that leads us to all sorts of faulty reasoning about knowledge. Faulty ideas about knowledge are troublesome at the best of times, but they're deadly to any country trying to make a transition to a "knowledge economy."

Fundamentally, the stuff we call "intellectual property" is just *knowledge*—ideas, words, tunes, blueprints, identifiers, secrets, databases. This stuff is similar to property in some ways: it can be valuable, and sometimes you need to invest a lot of money and labour into its development to realise that value.

Out of control

But it is also dissimilar from property in equally important ways. Most of all, it is not inherently "exclusive." If you trespass on my flat, I can throw you out (exclude you from my home). If you steal my car, I can take it back (exclude you from my car). But once you know my song, once you read my book, once you see my movie, it leaves my control. Short of a round of electroconvulsive therapy, I can't get you to un-know the sentences you've just read here.

It's this disconnect that makes the "property" in intellectual property so troublesome. If everyone who came over to my flat physically took a piece of it away with them, it'd drive me bonkers. I'd spend all my time worrying about who crossed the threshold, I'd make them sign all kinds of invasive agreements before they got to use the loo, and so on. And as anyone who has bought a DVD and been forced to sit through an insulting, cack-handed "You wouldn't steal a car" short film knows, this is exactly the kind of behaviour that property talk inspires when it comes to knowledge.

But there's plenty of stuff out there that's valuable even though it's not property. For example, my daughter was born on February 3, 2008. She's not my property. But she's worth quite a lot to me. If you took her from me, the crime wouldn't be "theft." If you injured her, it wouldn't be "trespass to chattels." We have an entire vocabulary and set of legal concepts to deal with the value that a human life embodies.

What's more, even though she's not my property, I still have a legally recognised *interest* in my daughter. She's "mine" in some meaningful sense, but she also falls under the purview of many other entities—the governments of the UK and Canada, the NHS, child protection services, even her extended family—they can all lay a claim to some interest in the disposition, treatment, and future of my daughter.

Flexibility and nuance

Trying to shoehorn knowledge into the "property" metaphor leaves us without the flexibility and nuance that a true knowledge rights regime would have. For example, facts are not copyrightable, so no one can be said to "own" your address, National Insurance Number, or the PIN for your ATM card. Nevertheless, these are all things that you have a strong interest in, and that interest can and should be protected by law.

There are plenty of creations and facts that fall outside the scope of copyright, trademark, patent, and the other rights that make up the hydra of Intellectual Property, from recipes to phone books to "illegal art" like musical mashups. These works are not property—and shouldn't be treated as such—but for every one of them, there's an entire ecosystem of people with a legitimate interest in them.

I once heard the WIPO representative for the European

association of commercial broadcasters explain that, given all the investment his members had put into recording the ceremony on the 60th anniversary of the Dieppe Raid in the second world war, they should be given the right to own the ceremony, just as they would own a teleplay or any other "creative work." I immediately asked why the "owners" should be some rich guys with cameras—why not the families of the people who died on the beach? Why not the people who own the beach? Why not the generals who ordered the raid? When it comes to knowledge, "ownership" just doesn't make sense—lots of people have an *interest* in the footage of the Dieppe commemoration, but to argue that anyone "owns" it is just nonsensical.

Copyright—with all its quirks, exceptions, and carve outs—was, for centuries, a legal regime that attempted to address the unique characteristics of knowledge, rather than pretending to be just another set of rules for the governance of property. The legacy of 40 years of "property talk" is an endless war between intractable positions of ownership, theft, and fair dealing.

If we're going to achieve a lasting peace in the knowledge wars, it's time to set property aside, time to start recognising that knowledge—valuable, precious, expensive knowledge—isn't owned. Can't be owned. The state should regulate our relative interests in the ephemeral realm of thought, but that regulation *must* be about knowledge, not a clumsy remake of the property system.

Saying Information Wants to Be Free Does More Harm Than Good

It's better to stop surveillance control because it is the people who really want to be free

For 10 years I've been part of what the record and film industry invariably call the "information wants to be free" crowd. In all that time, I've never heard anyone—apart from an entertainment executive—use that timeworn cliche.

"Information wants to be free" (IWTBF hereafter) is half of Stewart Brand's famous aphorism, first uttered at the Hackers Conference in Marin County, California (where else?), in 1984: "On the one hand information wants to be expensive, because it's so valuable. The right information in the right place just changes your life. On the other hand, information wants to be free, because the cost of getting it out is getting lower and lower all the time. So you have these two fighting against each other."

This is a chunky, chewy little koan, and as these go, it's an elegant statement of the main contradiction of life in the "information age." It means, fundamentally, that the increase in information's role as an accelerant and source of value is accompanied by a paradoxical increase in the cost of preventing the spread of information. That is, the more IT you have, the more IT generates value, and the more information becomes the centre of your world. But the more IT (and IT expertise) you have, the easier it is for information to spread and escape any proprietary barrier. As an oracular utterance predicting the next 40 years' worth of policy, business, and political fights, you can hardly do better.

But it's time for it to die.

It's time for IWTBF to die because it's become the easiest, laziest straw man for Hollywood's authoritarian bullies to throw up as a justification for the monotonic increase of surveillance, control, and censorship in our networks and tools. I can imagine them saying: "These people only want network freedom because they believe that 'information wants to be free.' They pretend to be concerned about freedom, but the only 'free' they care about is 'free of charge.'"

But this is just wrong. "Information wants to be free" has the same relationship to the digital rights movement that "kill whitey" has to the racial equality movement: a thoughtless caricature that replaces a nuanced, principled stand with a cartoon character. Calling IWTBF the ideological basis of the movement is like characterising bra burning as the primary preoccupation of feminists (in reality, the number of bras burned by feminists in the history of the struggle for gender equality appears to be zero, or as close to it as makes no difference).

So what do digital rights activists want, if not "free information"?

They want open access to the data and media produced at public expense, because this makes better science, better knowledge, and better culture—and because they already paid for it with their tax and licence fees.

They want to be able to quote, cite, and reference

earlier works because this is fundamental to all critical discourse.

They want to be able to build on earlier creative works in order to create new, original works because this is the basis of all creativity, and every work they wish to make fragmentary or inspirational use of was, in turn, compiled from the works that went before it.

They want to be able to use the network and their computers without mandatory surveillance and spyware installed under the rubric of "stopping piracy" because censorship and surveillance are themselves corrosive to free thought, intellectual curiosity, and an open and fair society.

They want their networks to be free from greedy corporate tampering by telecom giants that wish to sell access to their customers to entertainment congloms, because when you pay for a network connection, you're paying to have the bits you want delivered to you as fast as possible, even if the providers of those bits don't want to bribe your ISP.

They want the freedom to build and use tools that allow for the sharing of information and the creation of communities because this is the key to all collaboration and collective action—even if some minority of users of these tools use them to take pop songs without paying.

IWTBF has an elegant compactness and a mischievous play on the double-meaning of "free," but it does more harm than good these days.

Better to say, "The internet wants to be free."

Or, more simply: "People want to be free."

Chris Anderson's *Free* Adds Much to *The Long Tail*, but Falls Short

The economics of "free" goods and services cannot be explained in terms of the marketplace, digital or otherwise—humans are more complicated than that

This month saw the publication of the *Wired* U.S. editor-in-chief Chris Anderson's latest business book, *Free: The Future of a Radical Price*, a followup (of sorts) to the 2006 bestseller *The Long Tail*. I quite enjoyed *The Long Tail*, a book about the market opportunities created by the plummeting cost of inventory epitomized by the Amazons of the world.

While a traditional bookstore may stock a few thousand titles, Amazon can afford to "stock" (that is, list) millions of titles, and when they do so, they discover a remarkable thing: the titles that some bookstores ignored for absence of demand are, in fact, in demand. Not much demand—a book may sell a copy a year, or twice a decade—but where the cost of supplying that demand is nearly zero (Amazon's warehouse space is cheaper than a bookseller's retail shelf, and many of the books that Amazon sells are directly supplied by their publisher, or, increasingly, printed to order), it becomes possible to fulfil that demand.

The Long Tail resonated with me as a reader, a writer, and a former bookseller. As a reader, I knew that the books I loved were often nowhere to be found on the shelves of my local bookshop—not even in the so-called megastores that replaced the miserable mall stores that, in turn, had replaced the charming mom-and-pop stores. As a writer, I knew that once sales of my books had fallen off from their initial launch, the number of stores that carried them dropped off precipitously. And as a bookseller I knew that every day saw one or more customers looking for a book that we didn't carry—but always a different book.

But *The Long Tail* wasn't perfect. One area where I took great exception with its argument was when it came to digitally delivered goods, such as digital music, games, and books. Anderson's equation looks like this: [Goods]/[Cost of Inventory] = [Breadth of Market]. As [Cost of Inventory] fell, the market got bigger and more vibrant. It therefore followed that when [Cost of Inventory] fell to zero—as with the iTunes store, where the cost of running a store with 1,000,000 songs or 1,000,001 songs is, practically speaking, the same—the breadth of the market would be explosive.

But as every good programmer knows, dividing a number by zero yields an indeterminate outcome, and therein lay the problem with the hypothesis. As good as *The Long Tail* was at describing many kinds of markets, it didn't capture the extraordinary stuff that happened when the marginal cost of goods fell to zero.

For one thing, the cost of excluding people from those goods goes to infinity. Exclusion costs are a necessary part of any merchant's pricing model: a small newsagent's stall can set out piles of newspapers with saucers for coins on top, and use a hawk-eye and the social contract to stop people from walking away without paying. This lowers the newsagent's costs and increases his margins.

The cost of excluding people from commercially available digital goods is now infinite; this is another way of saying: "Any popular song, book, movie, TV show, or game will eventually be pirated." The only way to prevent this is to go to the impossible step of forcing everyone to trade in their PCs for specialised anti-copying devices, dismantling the internet as you do so. Failing that, exclusion is a lost cost.

Now, there's still a big market for non-excludable goods—whether it's the banana that sells at the cafe for eight times what it sells for at the grocer's next door, or the bottled water that you buy for several thousand times what it would cost you at your kitchen sink. But these aren't really goods in the way that, say, CDs or books or shirts are goods—they're services, the service being the convenience of getting them at this particular time with a minimum of hassle and fuss, at a price low enough not to bother about.

Likewise, iTunes sells a lot of music that you can get for free on the internet, so they're not really selling the music, they're selling the service of getting the music without having to muck about with P2P software and unsure quality.

Goods markets and service markets have very different characteristics, and *The Long Tail*'s lessons for digital service providers are necessarily different from the lessons it offers to those who use digital technology to improve the market for physical goods.

When I read *The Long Tail*, I thought Anderson had either run out of courage or vision when it came to digital information—the courage to consider that the market didn't explain, produce, or allocate the signature "product" of the 21st century; the vision to imagine what businesses centered on the service of "getting information more easily than you can get it elsewhere" might look like.

Enter *Free*, a book about the latter, but not the former. *Free* does a genuinely excellent job of describing the proven and speculative market opportunities that can be built around digital information services, from the musicians who use free downloads to fund paid gigs to the giant search companies that use free search to improve the market for paid advertising.

Some, such as Malcolm Gladwell, have faulted Anderson for failing to be sceptical enough of the businesses enabled by free, pointing out that services such as YouTube lack any sustainable revenue model (something that Anderson states in *Free*, contrasting it with its rival Hulu and making some shrewd observations about the potential future for both). Gladwell's criticisms ring hollow to me, blending a hand-wringing grievance about "theft" of information with special pleading for Gladwell and his fellow journalists.

Which is not to say that *Free* is perfect. Indeed, I think it has exactly the same problem as *The Long Tail*, namely, an unwillingness to consider the wider implications of a world centered on a commodity that can be infinitely reproduced at no marginal cost.

Nowhere is this more evident than in Anderson's dismissal of the Free Software Foundation founder, Richard Stallman, the original free software hacker who launched the GNU/Linux project that is the forebear of today's free/open source movement. Anderson mentions Stallman, dismissing him as "anti-capitalist."

But this is to miss one of the most important points. There's a pretty strong case to be made that "free" has some inherent antipathy to capitalism. That is, information that can be freely reproduced at no marginal cost may not want, need, or benefit from markets as a way of organising them.

And why not? There's plenty in our world that lives outside of the marketplace: it's a rare family that uses spot-auctions to determine the dinner menu or where to go for holidays. Who gets which chair and desk at your office is more likely to be determined on the lines of "from each according to his ability, to each according to his need" than on the basis of the infallible wisdom of the marketplace. The internally socialistic, externally capitalistic character of most of our institutions tells us that there's something to the idea that markets may not be the solution to all our problems.

And here's where *Free* starts to trip up. Though Anderson celebrates the best of non-commercial and anti-commercial net-culture, from amateur creativity to Freecycle, he also goes through a series of tortured (and ultimately less than convincing) exercises to put a dollar value on this activity, to explain the monetary worth of Wikipedia, for example.

And there is certainly some portion of this "free" activity that was created in a bid to join the non-free economy: would-be Hollywood auteurs who hope to be discovered on YouTube, for example. There's also plenty of blended free and non-free activity.

But for the sizeable fraction of this material—and it is sizeable—that was created with no expectation of joining the monetary economy, with no expectation of winning some future benefit for its author, that was created for joy, or love, or compulsion, or conversation, it is just wrong to say that the "price" of the material is "free."

The material, is, instead, literally priceless. It represents a large and increasing segment of our public life that is conducted entirely for reasons outside the marketplace. Some of the supporting planks may be market-driven (YouTube's free hosting), other parts are philanthropic (archive.org's free hosting), or simply so cheap that creators don't even notice the cost (any one of the many super-cheap hosting sites).

Through most of the history of the industrial era, markets were seen as a fit tool for organising a small piece of human endeavour, while the rest of life—the military, volunteerism, families, public service—were outside the marketplace. Markets may be good at organising scarce goods, and they may even be good at organising abundant ones, but do abundant goods really need organising?

Also missing in *Free* is the frank admission that for many of the practitioners threatened by digital technology, the future is bleak.

For while it is true that Madonna and many other established artists have found a future that embraces copying, there will also be many writers, musicians, actors, directors, game designers, and others for whom the internet will probably spell doom. And for every creator who loses her livelihood because she is unsuited to the digital future, there will be many more intermediaries—editors, executives, salespeople, clerks, engineers, teamsters, and printers—who will also be rendered jobless by technology.

It is possible to be compassionate about those peoples' fortunes—just as it is possible to mourn the passing of mom-and-pop bookstores, the collapse of poetry as a viable commercial concern, the worldwide decline of radio serials, the waning of the knife-sharpening trade, and a million other bygone human activities—while still not apologising for the future.

Anderson paints a rosy picture of "free," even noting the gains we all experienced as a result of the creative destruction of travel agents and stockbrokers thanks to Expedia and E*TRADE, but he fails to clearly and explicitly state something to the effect of: "The information revolution is not painless or bloodless. Its wrenching changes have and will put those of the industrial revolution to shame. Much of value will be lost."

On those lines, *Free* suffers from the same fate as many other recent business books: it describes a business-climate that no longer exists. The anecdotes and evidence come largely from the era of the cheap money bubble. Though there is a coda in which Anderson tries to sum up the lessons of *Free* for the econopocalypse, he fails to note with brutal honesty the fact that both of the free bubbles—dotcom and cheap money—had the side-effect of funding much of free's underpinnings, first by training millions of slacker undergrads in the basics of HTML and Perl at the expense of insurance-company-funded venture capitalists, and then by subsidising millions of experimental small "free" ventures as an indirect effect of over-capitalised advertisers pursuing a beggar-your-neighbour marketing strategy.

Indeed, there's something eerily Marxist in this phenomenon, in that it mirrors Marx's prediction of capitalism's ability to create a surplus of capacity that can subsequently be freely shared without market forces' brutality.

I'm not saying that "free" is communist, or even inherently anti-capitalist. But to discuss "free" without taking note of the ways in which it both challenges and reinforces non-market ways of living just as much as it does for market-driven ones is to only tell half the story.

Why Economics Condemns 3D to Be No More Than a Blockbuster Gimmick

You can't really *make a 3D movie while the money comes from 2D DVDs. And as for art-house 3D? Forget it*

My wife and I had a baby 18 months ago, which, practically speaking, means we've taken a year and more off from going to the cinema regularly, and only just started to get our heads of the water and get down to the movies.

Somewhere in the past year or so, it seems as though every studio exec has decided to greenlight one or more blockbuster in 3D, using a pretty impressive technology that employs polarised glasses that give a reasonably convincing illusion of depth. I have astygmatisms in each eye that make it difficult for me to converge most 3D, but I find I can get a pretty good effect with a minimum of (literal) headache if I sit in the centre of the back row.

And the 3D is...nice. Neil Gaiman's remarkable *Coraline* is thankfully devoid of the gimmicky 3D effects that characterized the last couple waves of 3D filmmaking. No viscera skewered on pikes hovering inches from your nose, no gag cans of spring-snakes leaping off the screen.

Just some lovely, quiet enhancements that are nice to have in a movie that is pretty fine to begin with.

But I'm sceptical.

Here's why: I just saw *Up*, the new Pixar movie, which is nearing the end of its run in Canada (the movie doesn't open in the UK until Christmas, but it's been playing in

North America for months now). *Up* is a tremendous movie, made me laugh and cry, and it was intended to be seen in 3D. (Pixar has the luxury of making its computer-rendered movies 3D simply by re-rendering them to produce the desired 3D effects.)

Because *Up* has been out in Canada for so long, it's been moved out of the rare 3D auditorium and into a regular screening room. And it's just fine, even without the 3D. Not for one second did I think "Oh, what I must be missing! If only I'd seen this in 3D!" Nothing was obviously missing from the 2D experience that made me feel like the 3D was a must-have.

And of course that's true of all 3D movies. Movies, after all, rely on the aftermarket of satellite, broadcast, and cable licenses, of home DVD releases and releases to airline entertainment systems and hotel room video-on-demand services, none of which are in 3D. If the movie couldn't be properly enjoyed in boring old 2D, the economics of filmmaking would collapse. So no filmmaker can afford to make a big-budget movie that is intended as a 3D-only experience, except as a vanity project.

What's more, no filmmaker can afford to make a *small*-budget 3D movie, either, because the cinema-owners who've shelled out big money to retrofit their auditoriums for 3D projection don't want to tie up their small supply of 3D screens with art-house movies. They especially don't want to do this when there's plenty of competition from giant-budget 3D movies that add in the 3D as an optional adjunct, a marketing gimmick that can be used to draw in a few more punters during the cinematic exhibition window.

I have no doubt that there are brilliant 3D movies lurking *in potentia* out there in the breasts of filmmakers, yearning to burst free. But I strongly doubt that any of them *will* burst free. The economics just don't support it: a truly 3D movie would be one where the 3D was so integral to the storytelling and the visuals and the experience that seeing it in 2D would be like seeing a giant-robots-throwing-buildings-at-each-other blockbuster as a flipbook while a hyperactive eight-year-old supplied the sound effects by shouting "BANG!" and "CRASH!" in your ear.

Such a film would be expensive to produce and market and could never hope to recoup. It won't be made. If it were made, it would not be followed.

In 10 years, we'll look back on the current round of 3D films and say, "Remember that 3D gimmick? Whatever happened to that, anyway? Hey, giant robot, watch where you're throwing that building!"

Not Every Cloud Has a Silver Lining

There's something you won't see mentioned by too many advocates of cloud computing—the main attraction is making money from you

The tech press is full of people who want to tell you how completely awesome life is going to be when everything moves to "the cloud"—that is, when all your important storage, processing, and other needs are handled by vast, professionally managed data-centres.

Here's something you won't see mentioned, though: the main attraction of the cloud to investors and entrepreneurs is the idea of making money from you, on a recurring, perpetual basis, for something you currently get for a flat rate or for free without having to give up the money or privacy that cloud companies hope to leverage into fortunes.

Since the rise of the commercial, civilian internet, investors have dreamed of a return to the high-profitability monopoly telecoms world that the hyper-competitive net annihilated. Investors loved its pay-per-minute model, a model that charged extra for every single "service," including trivialities such as Caller ID—remember when you had to pay extra to find out who was calling you? Imagine if your ISP tried to charge you for seeing the "FROM" line on your emails before you opened them! Minitel, AOL, MSN—these all shared the model, and had an iPhone-like monopoly over who could provide services on their networks, and what those service-providers would have to pay to supply these services to you, the user.

But with the rise of the net—the public internet, on which anyone could create a new service, protocol, or application—there was always someone ready to eat into this profitable little conspiracy. The first online services charged you for every email you sent or received. The next generation kicked their asses by offering email flat-rate. Bit by bit, the competition killed the meter running on your network session, the meter that turned over every time you clicked the mouse. Cloud services can reverse that, at least in part. Rather than buying a hard drive once and paying nothing—apart from the electricity bill—to run it, you can buy cloud storage and pay for those sectors every month. Rather than buying a high-powered CPU and computing on that, you can move your computing needs to the cloud and pay for every cycle you eat.

Now, this makes sense for some limited applications. If you're supplying a service to the public, having a cloud's worth of on-demand storage and hosting is great news. Many companies, such as Twitter, have found that it's more cost-effective to buy barrel-loads of storage, bandwidth, and computation from distant hosting companies than it would be to buy their own servers and racks at a data-centre. And if you're doing supercomputing applications, then tapping into the high-performance computing grid run by the world's physics centres is a good trick.

But for the average punter, cloud computing is—to say the least—oversold. Network access remains slower, more expensive, and less reliable than hard drives and CPUs. Your access to the net grows more and more fraught each day, as entertainment companies, spyware creeps, botnet crooks, snooping coppers, and shameless bosses arrogate to themselves the right to spy on, tamper with, or terminate your access to the net.

Alas, this situation isn't likely to change any time soon. Going into the hard-drive business or the computer business isn't cheap by any means—even with a "cloud" of Chinese manufacturers who'll build to your spec—but it's vastly cheaper than it is to start an ISP. Running a wire into the cellar of every house in an entire nation is a big job, and that's why you're lucky if your local market sports two or three competing ISPs, and why you can buy 30 kinds of hard drive on Amazon. It's inconceivable to me that network access will ever overtake CPU or hard drive for cost, reliability, and performance. Today, you can buy a terabyte of storage for £57. Unless you're recording hundreds of hours' worth of telly, you'd be hard-pressed to fill such a drive.

Likewise, you can buy a no-name quad-core PC with the aforementioned terabyte disc for £348. This machine will compute all the spreadsheets you ever need to tot up without breaking a sweat.

It's easy to think of some extremely specialised collaborative environments that benefit from cloud computing—we used a Google spreadsheet to plan our wedding list and a Google calendar to coordinate with my parents in Canada—but if you were designing these applications to provide maximum utility for their users (instead of maximum business-model for their developers), they'd just be a place where encrypted bits of state information was held for periodic access by powerful PCs that did the bulk of their calculations locally.

That's how I use Amazon's S3 cloud storage: not as an unreliable and slow hard drive, but as a store for encrypted backups of my critical files, which are written to S3 using the JungleDisk tool. This is cheaper and better than anything I could do for myself by way of offsite secure backup, but I'm not going to be working off S3 any time soon.

Why I Won't Buy an iPad (and Think You Shouldn't, Either)

I've spent ten years now on Boing Boing, finding cool things that people have done and made and writing about them. Most of the really exciting stuff hasn't come from big corporations with enormous budgets, it's come from experimentalist amateurs. These people were able to make stuff and put it in the public's eye and even sell it without having to submit to the whims of a single company that had declared itself gatekeeper for your phone and other personal technology.

Danny O'Brien does a very good job of explaining why I'm completely uninterested in buying an iPad—it really feels like the second coming of the CD-ROM "revolution" in which "content" people proclaimed that they were going to remake media by producing expensive (to make and to buy) products. I was a CD-ROM programmer at the start of my tech career, and I felt that excitement, too, and lived through it to see how wrong I was, how open platforms and experimental amateurs would eventually beat out the spendy, slick pros.

I remember the early days of the web—and the last days of CD-ROM—when there was this mainstream consensus that the web and PCs were too durned geeky and difficult and unpredictable for "my mom" (it's amazing how many tech people have an incredibly low opinion of their mothers). If I had a share of AOL for every time someone told me that the web would die because AOL was so easy and the web was full of garbage,

I'd have a lot of AOL shares.

And they wouldn't be worth much.

Incumbents made bad revolutionaries

Relying on incumbents to produce your revolutions is not a good strategy. They're apt to take all the stuff that makes their products great and try to use technology to charge you extra for it, or prohibit it altogether.

I mean, look at that Marvel app (just look at it). I was a comic-book kid, and I'm a comic-book grownup, and the thing that *made* comics for me was sharing them. If there was ever a medium that relied on kids swapping their purchases around to build an audience, it was comics. And the used market for comics! It was—and is—huge, and vital. I can't even count how many times I've gone spelunking in the used comic-bins at a great and musty store to find back issues that I'd missed, or sample new titles on the cheap. (It's part of a multigenerational tradition in my family—my mom's father used to take her and her sibs down to Dragon Lady Comics on Queen Street in Toronto every weekend to swap their old comics for credit and get new ones).

So what does Marvel do to "enhance" its comics? They take away the right to give, sell, or loan your comics. What an improvement. Way to take the joyous, marvellous sharing and bonding experience of comic reading and turn it into a passive, lonely undertaking that isolates, rather than unites. Nice one, Misney.

Infantalizing hardware

Then there's the device itself: clearly there's a lot of thoughtfulness and smarts that went into the design. But there's also a palpable contempt for the owner. I believe—really believe—in the stirring words of the Maker Manifesto: if you can't open it, you don't own it. Screws not glue. The original Apple II+ came with *schematics* for the circuit boards, and birthed a generation of hardware and software hackers who upended the world for the better. If you wanted your kid to grow up to be confident, entrepreneurial, and firmly in the camp that believes that you should forever be rearranging the world to make it better, you bought her an Apple II+.

But with the iPad, it seems like Apple's model customer is that same stupid stereotype of a technophobic, timid, scatterbrained mother as appears in a billion renditions of "that's too complicated for my mom" (listen to the pundits extol the virtues of the iPad and time how long it takes for them to explain that here, finally, is something that isn't too complicated for their poor old mothers).

The model of interaction with the iPad is to be a "consumer," what William Gibson memorably described as "something the size of a baby hippo, the color of a week-old boiled potato, that lives by itself, in the dark, in a double-wide on the outskirts of Topeka. It's covered with eyes and it sweats constantly. The sweat runs into those eyes and makes them sting. It has no mouth... no genitals, and can only express its mute extremes of murderous rage and infantile desire by changing the channels on a universal remote."

The way you improve your iPad isn't to figure out how it works and making it better. The way you improve the iPad is to buy iApps. Buying an iPad for your kids isn't a means of jump-starting the realization that the world is yours to take apart and reassemble; it's a way of telling your offspring that even changing the batteries is something you have to leave to the professionals.

Dale Dougherty's piece on Hypercard and its influence on a generation of young hackers is a must-read on this. I got my start as a Hypercard programmer, and it was Hypercard's gentle and intuitive introduction to the idea of remaking the world that made me consider a career in computers.

Wal-Martization of the software channel

And let's look at the iStore. For a company whose CEO professes a hatred of DRM, Apple sure has made DRM its alpha and omega. Having gotten into business with the two industries that most believe that you shouldn't be able to modify your hardware, load your own software on it, write software for it, override instructions given to it by the mothership (the entertainment industry and the phone companies), Apple has defined its business around these principles. It uses DRM to control what can run on your devices, which means that Apple's customers can't take their "iContent" with them to competing devices, and Apple developers can't sell on their own terms.

The iStore lock-in doesn't make life better for Apple's customers or Apple's developers. As an adult, I want to be able to choose whose stuff I buy and whom I trust to evaluate that stuff. I don't want my universe of apps constrained to the stuff that the Cupertino Politburo decides to allow for its platform. And as a copyright holder and creator, I don't want a single, Wal-Mart-like channel that controls access to my audience and dictates what is and is not acceptable material for me to create. The last time I posted about this, we got a string of apologies for Apple's abusive contractual terms for developers, but the best one was, "Did you think that access to a platform where you can make a fortune would come without strings attached?" I read it in Don Corleone's voice and it sounded just right. Of *course* I believe in a market where competition can take place without bending my knee to a company that has erected a drawbridge between me and my customers!

Journalism is looking for a daddy figure

I think that the press has been all over the iPad because Apple puts on a good show, and because everyone in journalism-land is looking for a daddy figure who'll promise them that their audience will go back to paying for their stuff. The reason people have stopped paying for a lot of "content" isn't just that they can get it for free, though: *it's that they can get lots of competing stuff for free,* too. The open platform has allowed for an explosion of new material, some of it rough-hewn, some of it slick as the pros, most of it targeted more narrowly than the old media ever managed. Rupert Murdoch can rattle his saber all he likes about taking his content out of Google, but I say *do it, Rupert.* We'll miss your fraction of a fraction of a fraction of a percent of the Web so little that we'll hardly notice it, and we'll have no trouble finding material to fill the void.

Just like the gadget press is full of devices that gadget bloggers need (and that no one else cares about), the mainstream press is full of stories that affirm the internal media consensus. Yesterday's empires do something sacred and vital and most of all *grown up,* and that other adults will eventually come along to move us all away from the kids' playground that is the wild web, with its amateur content and lack of proprietary channels where exclusive deals can be made. We'll move back into the walled gardens that best return shareholder value to the investors who haven't updated their portfolios since before E*TRADE came online.

But the real economics of iPad publishing tell a different story: even a stellar iPad sales performance isn't going to do much to stanch the bleeding from traditional publishing. Wishful thinking and a nostalgia for the good old days of lockdown won't bring customers back through the door.

Gadgets come and gadgets go

Gadgets come and gadgets go. The iPad you buy today will be e-waste in a year or two (less, if you decide not to pay to have the battery changed for you). The real issue isn't the capabilities of the piece of plastic you unwrap today, but the technical and social infrastructure that accompanies it.

If you want to live in the creative universe where anyone with a cool idea can make it and give it to you to run on your hardware, the iPad isn't for you.

If you want to live in the fair world where you get to keep (or give away) the stuff you buy, the iPad isn't for you.

If you want to write code for a platform where the only thing that determines whether you're going to succeed with it is whether your audience loves it, the iPad isn't for you.

Can You Survive a Benevolent Dictatorship?

The press loves the iPad, but beware Apple's attempt to shackle your readers to its hardware

The first press accounts of the Apple iPad have been long on emotional raves about its beauty and ease of use, but have glossed over its competitive characteristics—or rather, its lack thereof. Some have characterized the iPad as an evolution from flexible-but-complicated computers to simple, elegant appliances. But has there ever been an "appliance" with the kind of competitive control Apple now enjoys over the iPad? The iPad's DRM restrictions mean that Apple has absolute dominion over who can run code on the device—and while that thin shellac of DRM will prove useless at things that matter to publishers, like preventing piracy, it is deadly effective in what matters to Apple: preventing competition.

Maybe the iPad will fizzle. After all, that's what has happened to every other tablet device so far. But if you're contemplating a program to sell your books, stories, or other content into the iPad channel with hopes of it becoming a major piece of your publishing business, you should take a step back and ask how your interests are served by Apple's shackling your readers to its hardware. The publishing world chaos that followed the bankruptcy of Advanced Marketing Group (and subsidiaries like Publishers Group West) showed what can happen when a single distributor locks up too much of the business.

Apple isn't just getting big, however; it's also availing itself of a poorly thought-out codicil of copyright law to lock your readers into its platform, limit innovation in the e-book realm, and ultimately reduce the competition to serve your customers.

Jailbreak

Here's what most mainstream press reports so far haven't told you. The iPad uses a DRM system called "code-signing" to limit which apps it can run. If the code that you load on your device isn't "signed," that is, approved by Apple, the iPad will not run it. If the idea of adding this DRM to the iPad is to protect the copyrights of the software authors, we can already declare the system an abject failure—independent developers cracked the system within 24 hours after the first iPad shipped, a very poor showing even in the technically absurd realm of DRM. Code-signing has also completely failed for iPhones, by the way, on which anyone who wants to run an unauthorized app can pretty easily "jailbreak" the phone and load one up.

But DRM isn't just a system for restricting copies. DRM enjoys an extraordinary legal privilege previously unseen in copyright law: the simple act of breaking DRM is illegal, even if you're not violating anyone's copyright. In other words, if you jailbreak your iPad for the purpose of running a perfectly legal app from someone other than Apple, you're still breaking the law. Even if you've never pirated a single app, nor violated a single copyright, if you're found guilty of removing an "effective means of access control," Apple can sue you into a smoking hole. That means that no one can truly compete with Apple to offer better iStores, or apps, with better terms that are more publisher- and reader-friendly. Needless to say, it is also against the law to distribute tools for the purpose of breaking DRM.

Think about what that kind of control means for the future of your e-books. Does the company that makes your toaster get to tell you whose bread you can buy? Your dishwasher can wash anyone's dishes, not just the ones sold by its manufacturer (who, by the way, takes a 30% cut along the way). What's more, you can invent cool new things to do with your dishwasher. For example, you can cook salmon in it without needing permission from the manufacturer (check out the *Surreal Gourmet* for how). And you can even sell your dishwasher salmon recipe without violating some obscure law that lets dishwasher manufacturers dictate how you can use your machine.

Some early reviews have compared the iPad to a TV, a more passive medium in contrast to the interactive PC. But even passive old TV benefited greatly from the absence of a DRM-style lockdown on its medium. No one needed a broadcaster's permission, for example, to invent cable TV. No one needed a cable operator's permission to invent the VCR. And, tellingly, Apple cofounder Steve Wozniak didn't need a TV manufacturer's permission to invent the Apple II +, which plugged into the back of any old TV set. Of course, cable operators were sued by broadcasters, and the VCR was the subject of an eight-year court battle to wipe it off the face of the Earth. But by any measure, TV has greatly benefited from this system of "adversarial innovation." TiVo and all its imitators and successors, including the Apple TV, are good recent examples.

But this is not what is happening in e-book publishing so far. Devices like the iPad and the Kindle are a wholly new kind of thing—they function like bookshelves that reject all books except those the manufacturer has blessed. Publishers today worry that retailers like Wal-Mart might control too much of their business—and rightly so. But imagine how much more precarious things would be if Wal-Mart sold bookcases that were programmed to do what the iPad and Kindle do—refuse to hold books bought in other stores, and by canceling Wal-Mart's account, your publishing house would lose access to any customer who didn't have the desire to throw out their Wal-Mart bookcases or Wal-Mart–approved books, or room to add another brand of bookcase.

Having too much of your business subject to the whim of a single retailer who is out for its own interests is a scary and precarious thing. Already, Apple's App Store has displayed the warning signs of a less-than-benevolent dictator. Its standard deal with developers was, until recently, a secret—that is, until NASA was forced to reveal the terms of its deal with Apple in the face of a Freedom of Information Act request. Now that we've seen the details of that deal, we see what it means to sell into a marketplace with only one distributor: developers are prohibited from selling their apps in competing stores; consumers are prohibited from "jailbreaking" any Apple product even for legal uses; Apple can kick your app out of its store at any time; and Apple's liability to you is capped at $50, no matter what the circumstances. Apple has also announced a ban on the use of "middleware" programming environments that let you develop simultaneously for multiple platforms, like Google's Android OS, the Nintendo WiiWare marketplace, and so on.

Apple will tell you that it needs its DRM lock-in to preserve the iPad's "elegance." But if somewhere in the iPad's system settings there was a button that said, "I am a grownup and would like to choose for myself which apps I run," and clicking on that button would allow you to buy e-books from competing stores, where exactly is the reduction in elegance there?

Apple will also tell you that there's competition for apps—that anyone can write an HTML5 app (the powerful, flexible next generation of the HTML language that Web pages are presently made from). That may be true, but not if developers want their app to access the iPad's sensors, which allow apps to be bought and sold with a single click. It's an enormous competitive setback if your customers have to laboriously tap their credit card details into the screen keyboard every time they buy one of your products. And here's a fun experiment for the code writers among you: write an app and stick a "buy in one click with Google Checkout" button on the screen. Watch how long it takes for Apple to reject it. For bonus fun, send the rejection letter to the FTC's competition bureau. Whaddya Gonna Do?

There's an easy way to change this, of course. Just tell Apple it can't license your copyrights—that is, your books—unless the company gives you the freedom to give your readers the freedom to take their products with them to any vendor's system. You'd never put up with these lockdown shenanigans from a hardcopy retailer or distributor, and you shouldn't take it from Apple, either, and that goes for Amazon and the Kindle, too.

This is exactly what I've done. I won't sell my e-books in any store that locks my users into a vendor's platform. That's true for both my forthcoming self-published collection *With a Little Help* and the e-book editions that HarperCollins and Tor publish of my books. At the same time, I'm hoping my unlocked readers will come up with great HTML5 remixes of the stories in *With a Little Help*: interactive, cross-platform net-toys that can actually drive revenue for me, whether through sales of my print editions, donations for the e-books, or downloads of the audio.

I'm planning to be in the publishing business for a good half-century or more. And though I am not exactly sure how the e-publishing book business will mature (hence my experiment *With a Little Help*), I am keenly aware that locking my readers to a specific device today, whether the iPad or the Kindle, could very well mean a dramatic loss of control for my business tomorrow.

Curated Computing Is No Substitute for the Personal and Handmade

Bespoke computing experiences promise a pipe dream of safety and beauty—but the real delight lies in making your own choices

The launch of the iPad and the general success of mobile device app stores has created a buzzword frenzy for "curated" computing—computing experiences where software and wallpaper and attendant foofaraw for your device are hand-picked for your pleasure.

In theory, this creates an aesthetically uniform, and above all *safe* and easy, computing environment, as the curators see to it that only the very prettiest, easiest-to-use, and most virus-free apps show up in the store.

I'm all for it. After all, I've spent the past 10 years co-curating Boing Boing, a place where my business-partners and I pick the websites that interest us the most and assemble them into a kind of deep, wide, searchable catalogue of things that you should know, do, and marvel at.

We've recently launched a store, the Boing Boing Bazaar, consisting of the most interesting inventions, clothing, gadgets, decor, and assorted gubbins that our readers have created, as picked by us. My Twitter account mostly consists of retweets from other twitterers—my collection of the best tweets I've seen today. I am a born curator, and have spent my life amassing collections and showing them off.

But there's something important to note about all

these curatorial roles I enjoy: none of them are coercive. No one forces you to read Boing Boing, and if you do, there's nothing that prevents you from reading another weblog (or a couple hundred other weblogs). Order as many gizmos as you'd like from the Boing Boing Bazaar, we'll never tell you that you can't fill your knick-knack shelves from anyone else's curated *wunderkammer*. Follow me on Twitter if it pleases you, and feel free to follow anyone else you find interesting.

The beauty of noncoercive curation is that there are so many reasons we value things, it's really impossible to imagine that any one place will serve as a one-stop shop for our needs.

Two categories in particular won't ever be fulfilled by a curator: first, the personal. No curator is likely to post pictures of my family, videos of my daughter, notes from my wife, stories I wrote in my adolescence that my mum's recovered from a carton in the basement.

My own mediascape includes lots of this stuff, and it is every bit as compelling and fulfilling as the slickest, most artistic works that show up in the professional streams. I don't care that the images are overexposed or badly framed, that the audio is poor quality, that I can barely read my 14-year-old self's handwriting. The things I made with my own hands and the things that represent my relationships with my community and loved ones are critical to my identity, and I won't trade them for anything.

Second, the tailored. I have loads of little scripts, programs, systems, files, and such that make perfect sense to me, even though they're far from elegant or perfect. There's the script I use for resizing and uploading images to Boing Boing, the shelf I use to organise my to-be-read pile, the carefully-built mail rules that filter out spam and trolls and make sure I see the important stuff. I am a market of one: no one wants to make a commercial proposition out of filling my needs, and if they did, your average curator would be nuts to put something so tightly optimised for my needs into the public sphere, where it would be so much clutter. But again, these are the nuts and bolts that hold my life together and I can't live without them.

In a noncoercive curatorial world, these categories can peacefully coexist with curated spaces. There are hundreds of places where I can find recommendations and lists and reviews and packages of software for my computer (Ubuntu, the version of GNU/Linux I use, has its own very good software store). I can use as many or as few of these curators as I'd like, and what's more, I can add in things that matter to me because they exactly suit my needs or fulfil some sentimental niche in my life.

But I fear that when analysts slaver over "curated" computing, it's because they mean "monopoly" computing—computing environments like the iPad where all your apps have to be pre-approved by a single curating entity, one who uses the excuse of safety and consistency to justify this outrageous power grab. Of course, these curators are neither a guarantee of safety, nor of quality: continuous revelations about malicious software and capricious, inconsistent criteria for evaluating software put the lie to this. Even without them, it's pretty implausible to think that an app store with hundreds of thousands (if not millions) of programs could be blindly trusted to be free from bugs, malware, and poor aesthetic choices.

No, the only real reason to adopt coercive curation is to attain a monopoly over a platform—to be able to shut out competitors, extract high rents on publishers whose materials are sold in your store, and sell a pipe dream of safety and beauty that you can't deliver, at the cost of homely, handmade, personal media that define us and fill us with delight.

Doctorow's First Law

With a Little Help is on track for a September release. The printer has found the right paper; the binder is ready to do a test binding; and all is well on Earth. I know, I said July—and I could have launched in July—but that would have meant interrupting the launch with a monthlong, internet-free family holiday in August, right around the time I get back from the World Science Fiction Convention in Melbourne, Australia, and my subsequent German and Dutch tours.

In the meantime, I've been filling the time productively by attempting to discover which online booksellers exist to serve the interests of copyright owners, like me, and which ones are seeking to unfairly bind copyright holders (and consumers) to their platforms and, as a result, diminish our negotiating power. I'm happy to say that after much work, I have persuaded three major retailers to offer my e-books without any technology or license conditions that would prohibit my customers from moving the e-books they've purchased to a competitor's device: Amazon, Barnes & Noble, and Kobo.

Lockdown

Strange as it may sound, this is a major victory. I first discovered just how little bargaining power creators have in the digital marketplace back in 2008, when Random House Audio and I approached Audible, the Amazon division that controls most of the audiobook market and the sole audiobook supplier to iTunes, and asked them to carry the audio edition of my *New York Times* bestseller *Little Brother* without DRM. We were turned down.

To Audible's credit, they relented when my next book, *Makers*, came out. However, they informed me that Apple, their largest retail partner, would not carry the title in the iTunes store without DRM. And despite agreeing to forgo the DRM, Audible's license agreement still contained contract prohibitions on customers' moving their property to competing platforms, along with plenty of other terms that either were no good for my business interests and/or were terms that I personally would not agree to in order to buy an audiobook.

I tried to remedy this creatively by asking Audible to allow me to add some preliminary language to the audio edition, something that said, "notwithstanding any agreement you clicked on to buy this book, Cory Doctorow and Random House Audio, as the copyright holders, hereby give you their blessing to do anything that is permitted by your local copyright law." In other words: don't break the law, but feel free to do anything else—the same terms under which your car, dishwasher, and every traditional book on your bookshelf was sold to you. Again, Audible declined.

This led me to formulate something I grandiosely call Doctorow's First Law: "Any time someone puts a lock on something that belongs to you, and won't give you a key, they're not doing it for your benefit."

This year, I set out to test this law. In May, I cornered Macmillan CEO John Sargent and CTO Fritz Foy at the Macmillan BEA party. As the publishers of my books with Tor, I asked them if they'd be willing to try offering my e-books to all the major online booksellers—Amazon's Kindle store, Apple's iPad store, Barnes & Noble's Nook store, Sony's e-book store, and Kobo—as DRM-free products with the following text inserted at the beginning of the file:

"If the seller of this electronic version has imposed contractual or technical restrictions on it such that you have difficulty reformatting or converting this book for use on another device or in another program, please visit http://craphound.com for alternate, open format versions, authorized by the copyright holder for this work, Cory Doctorow. While Cory Doctorow cannot release you from any contractual or other legal obligations to anyone else that you may have agreed to when purchasing this version, you have his blessing to do anything that is consistent with applicable copyright laws in your jurisdiction."

As I explained to John and Fritz, although all my books are available as downloads for free, I often hear from readers who want to buy them, either because it is a simple way to compensate me (I also maintain a public list of schools and libraries who've solicited copies of my books so that grateful e-book readers can purchase and send a print copy to one of them, thus repaying my favor and doing a good deed at the same time) or because they like the no-hassle option of tapping on their device to buy a book. I am more than happy to offer my otherwise free books for sale in any vendor's store, of course, but only if the vendors agree to carry them on terms I feel I can stand behind as an entrepreneur, as an artist, and as a moral actor.

John and Fritz strongly supported the idea. Macmillan, after all, had just gone to the mat with Amazon for control of e-book terms of sale, making control a priority in its future dealings with electronic retail and wholesale channels. Now, there are some writers, agents, or publishers that want DRM and restrictive EULAs. And though I can't understand why, we are at least in agreement on this point: it should be the copyright holder's choice. When it comes to which restrictions copyright law should place on e-book readers, the copyright proprietor—whether the author or the publisher—should call the shots, not the retailers.

I'm happy to report that Amazon, to its eternal credit, was delighted to offer my e-books without DRM and with the anti-EULA license language, as was Barnes & Noble and Kobo. Why Amazon's Kindle division was happy to do what its Audible division had categorically rejected is still beyond me, but I'll take any sign of fairness I can get. I can only hope that Amazon's other digital divisions catch up with Kindle, and if they do, I'll be eager to have my audiobooks for sale in the Audible store. Amazon is a retailer that has literally revolutionized my life, my go-to supplier for everything from toilet brushes to used DVDs for my toddler. And in addition to selling my own works, I also sell upwards of 25,000 books a year through Amazon affiliate links in my online book reviews. This makes me a one-man, good-sized independent bookstore, with Amazon doing my fulfillment, payment processing, stocking, etc.

Unfortunately, I had no such luck with Apple or Sony. True to my earlier experience with Apple's iTunes store, Apple has a mandatory DRM requirement for books offered for sale for the iPad. I know many Apple fans believe that because Steve Jobs penned an open letter decrying DRM that the company must use DRM because they have no choice. But this simply isn't true. Sony has the same deal.

Cracked Thinking

Dirty fighting instructors say: "any weapon you don't know how to use belongs to your enemy." One illustrative example of this principle is to be found in DRM. Until last week, U.S. law protected DRM, making it illegal to break encryption or other technological protections under nearly any circumstance. In other words, if Apple offers a DRM-locked edition of one of my books, even I am not legally allowed to remove the DRM without Apple's permission, even if I'm making a perfectly legal use under copyright law. And I certainly can't authorize my readers to do so.

On July 26, the law eased a little when the U.S. Copyright Office granted an exemption to the Digital Millennium Copyright Act that allows DRM-cracking on iPhones and other mobile phones (and possibly other devices, like iPads, though no one knows for sure) for the purpose of installing third-party software. But the exemption doesn't allow for the creation or distribution of tools to accomplish this, which makes the whole thing something of a Pyrrhic victory. And remember, digital editions are generally licensed, not owned. Therefore, just because you may not be breaking the law by cracking your device, if you're violating your license terms you can still be denied service.

If you think about it, this is a rather curious circumstance, because it means that once a technology company puts a lock on a copyrighted work, the proprietor of that copyright loses the right to authorize his audience to use it in new ways, including the right to authorize a reader to move a book from one platform to another. At that point, DRM and the laws that protect it stop protecting the wishes of creators and copyright owners, and instead protect the business interests of companies whose sole creative input may be limited to assembling a skinny piece of electronics in a Chinese sweatshop.

What's more, many of these distribution channels won't even allow copyright holders the option of presenting their works without DRM. So if you sell one million dollars worth of DRM-locked Kindle books, you are essentially a million bucks in hock to Amazon—that being the cost of the investment you'd have to ask your audience to abandon in order to switch to a competing platform. How does giving your retail partners that kind of market control benefit you, the copyright holder?

Still, I'm encouraged by the actions of Amazon, Barnes & Noble, and Kobo in adopting my terms for sale on their platforms. Hey, three out of five is a pretty good showing—it's three more than we had a year ago. And it gives me hope that authors and the publishing industry can pull together and start to demand that all the retail channels yield copyright control to creators and publishers, rather than hijacking it to their benefit.

Reports of Blogging's Death Have Been Greatly Exaggerated

Blogging is not on the way out—it's just that other social media have taken over many of its functions

A report last month in the *Economist* tells us that "blogging is dying" as more and more bloggers abandon the form for its cousins: the tweet, the Facebook Wall, the Digg.

Do a search-and-replace on "blog" and you could rewrite the coverage as evidence of the death of television, novels, short stories, poetry, live theatre, musicals, or any of the hundreds of the other media that went from breathless ascendancy to merely another tile in the mosaic.

Of course, none of those media are dead, and neither is blogging. Instead, what's happened is that they've been succeeded by new forms that share some of their characteristics, and these new forms have peeled away all the stories that suit them best.

When all we had was the stage, every performance was a play. When we got films, a great lot of these stories moved to the screen, where they'd always belonged (they'd been squeezed onto a stage because there was no alternative). When TV came along, those stories that were better suited to the small screen were peeled away from the cinema and relocated to the telly. When YouTube came along, it liberated all those stories that wanted to be 3–8 minutes long, not a 22-minute sitcom or a 48-minute drama. And so on.

What's left behind at each turn isn't less, but *more:* the stories we tell on the stage today are there not because they must be, but because they're better suited to the stage than they are to any other platform we know about. This is wonderful for all concerned—the audience numbers might be smaller, but the form is much, much better.

When blogging was the easiest, most prominent way to produce short, informal, thinking-aloud pieces for the net, we all blogged. Now that we have Twitter, social media platforms, and all the other tools that continue to emerge, many of us are finding that the material we used to save for our blogs has a better home somewhere else. And some of us are discovering that we weren't bloggers after all—but blogging was good enough until something more suited to us came along.

I still blog 10–15 items a day, just as I've done for 10 years now on Boing Boing. But I also tweet and retweet 30–50 times a day. Almost all of that material is stuff that wouldn't be a good fit for the blog—material I just wouldn't have published at all before Twitter came along. But a few of those tweets might have been stretched into a blogpost in years gone by, and now they can live as a short thought.

For me, the great attraction of all this is that preparing material for public consumption forces me to clarify it in my own mind. I don't really know it until I write it. Thus the more media I have at my disposal, the more ways there are for me to work out my own ideas.

Science fiction writer Bruce Sterling says: "The future

composts the past." There's even a law to describe this, Riepl's Law—which says "new, further developed types of media never replace the existing modes of media and their usage patterns. Instead, a convergence takes place in their field, leading to a different way and field of use for these older forms."

That was coined in 1913 by Wolfgang Riepl. It's as true now as it was then.

Streaming Will Never Stop Downloading

Far from being a cure for the industry's woes, substituting streams for downloads wastes bandwidth, reduces privacy, and slows innovation

Someone convinced the record and movie and TV industries that there is way of letting someone listen to audio or watch video over the internet without making a copy. They call this "streaming" audio, and compare it to radio, and contrast it with "downloading," which they compare to buying a CD.

The idea that you can show someone a movie over the internet without making a copy has got lots of people in policy circles excited, since it seems to "solve the copyright problem." If services such as Hulu, Last.fm, and YouTube can "play you a file" instead of "sending you a file," then we're safely back in the pre-Napster era. You can sell subscriptions to on-demand streaming, and be sure that your subscribers will never stop paying, since they don't own their favourite entertainment and will have to stump up in order to play it again.

There's only one problem: Streaming doesn't exist.

Oh, OK. Streaming *exists.* It is a subset of downloading, which comes in many flavours. Downloading is what happens when one computer (a server, say) sends another computer (your PC, say) a file. Some downloads happen over http, the protocol on which the web is based. Some happen over BitTorrent, which pulls the file from many different locations, in no particular order, and reassembles it on your side. Some downloads take place over secure protocols like SSH and SSL, and some are part of intelligent systems that, for example, keep your computer in sync with an encrypted remote backup.

Streaming describes a collection of downloading techniques in which the file is generally sent sequentially, so that it can be displayed before it is fully downloaded. Some streams are open-ended (like the video stream coming off your CCTV camera, which isn't a finite file, but rather continues to transmit for as long as the CCTV is up and running).

Some travel over UDP, a cousin of the more familiar TCP, in which reliability can be traded off for speed. Some streaming servers can communicate with the downloading software and dynamically adjust the stream to compensate for poor network conditions.

And of course, some streaming software throws away the bits after it finishes downloading them, rather than storing them on the hard drive.

It's this last part that has the technologically naive excited. They assume that because a downloading client can be designed in such a way that it doesn't save the file, no "copy" is being made. They assume that this is the technical equivalent of "showing" someone a movie instead of "giving them a copy" of it.

But the reason some download clients discards the bits is because the programmer chose not to save them. Designing a competing client that doesn't throw away the bits—one that "makes a copy"—is trivial.

All streaming involves making a copy, and saving the copy just isn't hard.

Does this matter? After all, if the entertainment industry can be bought off with some pretty stories about a magical kind of download that doesn't make a copy, shouldn't we just leave them to their illusions?

What harm could come from that?

Plenty, I fear. First of all, while streaming music from Last.fm is a great way to listen to music you haven't discovered yet, there's no reason to believe that people will lose the urge to collect music.

Indeed, the record industry seems to have forgotten the lesson of 70 years' worth of radio: people who hear songs they like often go on to acquire those songs for their personal collections. It's amazing to hear record industry executives deny that this will be the case, especially given that this was the dominant sales strategy for their industry for most of a century. Collecting is easier than it has ever been: you can store more music in less space and organise it more readily than ever before.

People will go on using streaming services, of course. They may even pay for them. But people will also go on downloading. Streaming won't decrease downloading. If streaming is successful—that is, if it succeeds in making music more important to more people—then downloading will increase too. With that increase will come a concomitant increase in Big Content's attacks on the privacy and due process rights of internet users, which, these days, is pretty much everyone.

If you want to solve the "downloading problem" you can't do it by waving your hands and declaring that a totally speculative, historically unprecedented shift in user behaviour—less downloading—will spontaneously arise through the good offices of Last.fm.

There are more problems, of course. Streaming is an implausible and inefficient use of wireless bandwidth. Our phones and personal devices can be equipped with all the storage necessary to carry around tens of thousands of songs for just a few pounds, incurring a single cost. By contrast, listening to music as you move around (another factor that has been key to the music industry's strategy, starting with the in-car eight-track player and continuing through the Walkman and iPod) via streams requires that you use the scarce electromagnetic spectrum that competing users are trying to get their email or web pages over. Count the number of earbuds on the next tube-carriage, airplane, or bus you ride, multiply it by 128kbps (for a poor quality audio stream) and imagine that you had to find enough wireless bandwidth to serve them all, without slowing down anyone's competing net applications. Someday, every 777 might come with a satellite link, but will it provide all 479 passengers with enough bandwidth to play music all the way from London to Sydney?

What's more, streaming requires that wireless companies be at the centre of our daily cultural lives. These are the same wireless companies that presently screw us in every conceivable way: charging a premium for dialling an 0870 number; having limits on "unlimited" data plans; charging extra for "long distance" text messages. They're the same wireless companies whose hold-queues, deceptive multi-year contracts, surprise bills, and flaky network coverage have caused more bad days than any other modern industry.

Why would we voluntarily increase our reliance on expensive, scarce wireless bandwidth delivered by abusive thugs when we are awash in cheap, commodity storage that grows cheaper every day and which we can buy from hundreds of manufacturers and thousands of retailers?

Especially when every streaming song creates a raft of privacy disclosures—your location, your taste, even the people who may be near you and when you're near them—that are far more controllable when you listen to your own music collection.

Finally, there's the cost of going along with the gag. The more we pretend that there is a technical possibility of designing a downloader that can't save its files, the more incentive we create for legal and technological systems that attempt to make this come true. The way you hinder a downloader from saving files is by obfuscating its design and by creating legal penalties for users who open up the programs they use and try to improve them. You can't ever have a free/open source downloader that satisfies the desire to enforce deletion of the file on receipt, because all it would take to remove this stricture is to modify the code.

An incentive to obfuscate code, to prohibit third-party modifications and improvements, and to weld the bonnet shut on all the world's computers won't actually stop downloading. But it will have anti-competitive effects, it will reduce privacy, and it will slow down innovation, by giving incumbents the right to control new entrants into the market.

Hard problems can't be solved with technical denialism. The market has spoken: people want to download their music (and sometimes they want to stream it, too). The supposedly for-profit record labels could offer all-you-can-download packages that captured the law-abiding downloader, and then they could retain those customers by continuing to make new, great music available. It's been 10 years since Napster, and the record industry's hypothesis that an all-you-can-download regime can't work because users will download every song and then unsubscribe from the service is not borne out by evidence. The fact is that most downloaders find cheap, low-risk music discovery to be a tremendous incentive to *more* consumption, as they discover new music, new artists, new songs, and new genres that tickle their fancies.

Selling customers what they desire is fundamental to any successful business. If Big Content can't figure out how to do that, then we can only pray for their hasty demise, before they can do too much more damage to humanity's most amazing and wonderful invention: the internet.

Search Is Too Important to Leave to One Company—Even Google

It may seem as unlikely as a publicly edited encyclopedia, but the internet needs publicly controlled search

Search is the beginning and the end of the internet. Before search, there was the idea of an organised, hierarchical internet, set up along the lines of the Dewey Decimal system.

Again and again, net pioneers tried to build such systems, but they were always outcompeted by the messy hairball of the real world. As Wikipedia shows, building consensus about what goes where in a big org chart is hard, and the broader the subject area, the harder it gets.

Melvin Dewey didn't predict computers; he also mixed Islam in with Sufism, and gave table-knocking psychics their own category. A full-contact sport like the internet just doesn't lend itself to a priori categorisation.

Enter search. Who needs categories, if you can just pile up all the world's knowledge every which way and use software to find the right document at just the right time?

But this is not without risk: search engines accumulate near-complete indexes of our interests, our loves, our hopes, and aspirations. Our relationship with them is as intimate as our relationships with our lovers, our confessors, our therapists.

What's more, the way that search engines determine

the ranking and relevance of any given website has become more critical than the editorial berth at the *New York Times* combined with the chief spots at the major TV networks. Good search engine placement is make-or-break advertising. It's ideological mindshare. It's relevance.

Contrariwise: being poorly ranked by a search engine makes you irrelevant, broke, and invisible.

What's more, search engines routinely disappear websites for violating unpublished, invisible rules. Many of these sites are spammers, link-farmers, malware sneezers, and other gamers of the system. That's not surprising: every complex ecosystem has its parasites, and the internet is as complex as they come. The stakes for search-engine placement are so high that it's inevitable that some people will try anything to get the right placement for their products, services, ideas, and agendas. Hence the search engine's prerogative of enforcing the death penalty on sites that undermine the quality of search.

It's a terrible idea to vest this much power with one company, even one as fun, user-centered, and technologically excellent as Google. It's too much power for a handful of companies to wield.

The question of what we can and can't see when we go hunting for answers demands a transparent, participatory solution. There's no dictator benevolent enough to entrust with the power to determine our political, commercial, social, and ideological agenda. This is one for The People.

Put that way, it's obvious: if search engines set the

public agenda, they should be public. What's not obvious is how to make such a thing.

We can imagine a public, open process to write search engine ranking systems, crawlers, and the other minutiae. But can an ad hoc group of net-heads marshall the server resources to store copies of the entire internet?

Could we build such a thing? It'd be as unlikely as a noncommercial, volunteer-written encyclopedia. It would require vast resources. But it would have one gigantic advantage over the proprietary search engines: rather than relying on weak "security through obscurity" to fight spammers, creeps, and parasites, such a system could exploit the powerful principles of peer review that are the gold standard in all other areas of information security.

Google itself was pretty damned unlikely—two grad students in a garage going up against vast, well-capitalised mature search companies like AltaVista (remember them?). Search is volatile and we'd be nuts to think that Google owned the last word in organising all human knowledge.

Copyright Enforcers Should Learn Lessons from the War on Spam

Those who forget history are doomed to repeat it.

For example: say you're an entertainment executive looking to stop some incredibly popular kind of online information transmission—infringing music copyright, say. Where would you look to find a rich history of this kind of online battle? Why, the Spam Wars, of course. Where else?

Electronic spam has existed in one form or another since 1978. For 30 years, networks have served as battlefield in the fight between those who want your mailbox filled with their adverts and those who want to help you avoid the come-ons.

The war against spam has been a dismal failure: there's far more spam today than ever before, and it grows more sinister by the day. Gary Thuerk's 1978 bulk email advertisement for a new Digital Equipment model (widely held to be the first spam) was merely annoying and gormless. Today, the spam you receive might hijack your computer, turning it into a spyware-riddled zombie that harvests your banking details and passwords and uses its idle resources to send out even more spam. It might encrypt your files and demand anonymous cash transfers before unlocking them. It might be a front for a Spanish Prisoner scammer who will rob you of every cent you and your loved ones have.

And (practically) everyone hates spam. It's not like copyrighted music, where millions of time-rich, cash-poor teenagers and cheapskates are willing to spend their days and nights figuring out how to get more of it in their lives. In the Spam War, the message recipients are enthusiastic supporters of the cause.

Let's have a look at some of the spam war tactics that have been tried and have been found wanting.

Content-based filters

These were pretty effective for a very brief period, but the spammers quickly outmanoeuvred them. The invention of word-salads (randomly cut/pasted statistically normal text harvested from the net), alphabetical substitutions, and other tricksy techniques have trumped the idea that you can fight spam just by prohibiting certain words, phrases, or media.

> **Unintended consequence:** It's practically impossible to have an email conversation about Viagra, inheritances, medical conditions related to genitals, and a host of other subjects because of all the "helpful" filters still fighting last year's spam battle, diligently vaporising anyone who uses the forbidden words.

Blacklisting

Anti-spam groups maintain blacklists of "rogue" internet service providers and their IPs—the numbers that identify individual computers. These are ISPs that, due to negligence, malice, error, or a difference of opinion on how to best block bad actors, end up emitting a lot of spam to the rest of the internet. Again, this worked pretty well for a short period, but was quickly overwhelmed by more sophisticated spammers who switched from running rogue email servers to simply hijacking end users' PCs and using them to send spams from millions of IPs.

> **Unintended consequence:** IP blocking becomes a form of collective punishment in which innocent people are punished (blocked from part or all of the internet) because one person did something naughty, and none of the punished had the power to prevent it. A single IP can stand in for thousands or even millions of users.
>
> The blacklists are maintained by groups whose identity is shrouded in secrecy ("to prevent retaliation from criminal spam syndicates") and operate at Star Chambers who convict their targets in secrecy, without the right of appeal or the ability to confront your accuser. Allegations abound that blacklisters have targeted their critics and stuck them in the black holes merely for criticising them, and not because of any spam.

Blocking open servers

Email servers used to be set up to accept and deliver mail for anyone: all you needed to do to send an email was to contact any known email server and ask it to forward your message for you. This made email sending incredibly easy to set up and run—if your local mailserver croaked, you could just switch to another one. But these servers were also juicy targets for spammers who abused their hospitality to send millions of spams. A combination of blacklisting and social pressure has all but killed the open server in the wild.

> **Unintended consequence:** It's infinitely harder to send legitimate email, as anyone who has ever logged into a hotel or institutional network and discovered that you can't reach your mailserver any more can attest. And still the spam rolls in: legitimate users lack the motivation and capacity to learn to send mail in a block-ridden environment, whereas spammers have the motivation and capacity in spades.

There have been other failures in the field, and a few successes (my daily spam influx dropped from more than 20,000 to a few hundred when my sysadmin switched on something called greylisting). But these three failures are particularly instructive because they represent the main strategic objectives of the entertainment industry's copyright enforcement plans.

Every legislative and normative proposal recapitulates the worst mistakes of the spamfight: from Viacom's demand that Google automatically detect copyright-infringing videos while they're being uploaded; to the three-accusations-and-you're-offline proposal from the BPI; to the notion in the G8's Anti-Counterfeiting Trade Agreement of turning copyright holders into judge, jury, and executioner for what content can travel online and who can see it.

The Spam Wars have shown us that great intentions and powerful weapons can have terrible outcomes—

outcomes where the innocent are inconvenienced and the guilty merely evolve into more resistant, more deadly organisms.

Warning to All Copyright Enforcers: Three Strikes and *You're* Out

I think we should permanently cut off the internet access of any company that sends out three erroneous copyright notices. Three strikes and you're out, mate.

Having been disconnected, your customers can only find out about your product offerings by ringing you up and asking, or by requesting a printed brochure. Perhaps you could give all your salespeople fax machines so they can fax urgent information up and down the supply chain. And there's always the phone—just make sure you've got a bunch of phone books in the office, because you'll never Google another phone number.

Call it a modest proposal in the Swiftian sense if you must, but I'm deadly serious.

You see, the big copyright companies—record labels, broadcasters, film studios, software companies—are lobbying in the halls of power around the world (including Westminster) for a three strikes rule for copyright infringers. They want to oblige internet service providers (ISPs) to sever the broadband links of any customer who has been thrice accused of downloading infringing material, and to oblige web-hosting companies to terminate the accounts of anyone accused of sticking infringing material on a web server three times.

They're not even proposing that this punishment should be reserved for *convicted* infringers. Proving infringement is slow and expensive—so much so that the Motion Picture Association of America just filed a brief with the U.S. court considering the appeal of Jammie Thomas, a woman sentenced to pay $222,000 in fines for downloading music, in which the trade association argued that they should *never* have to prove infringement to collect damages, since proof is so hard to come by.

I mean, it's not as though internet access is something *important*, right?

In the past week, I've only used the internet to contact my employers around the world, my MP in the UK, to participate in a European Commission expert proceeding, to find out why my infant daughter has broken out in tiny pink polka-dots, to communicate with a government whistle-blower who wants to know if I can help publish evidence of official corruption, to provide references for one former student (and follow-up advice to another), book my plane tickets, access my banking records, navigate the new Home Office immigration rules governing my visa, wire money to help pay for the headstone for my great uncle's grave in Russia, and to send several Father's Day cards (and receive some of my own).

The internet is only that wire that delivers freedom of speech, freedom of assembly, and freedom of the press in a single connection. It's only vital to the livelihood, social lives, health, civic engagement, education, and leisure of hundreds of millions of people (and growing every day).

This trivial bit of kit is so unimportant that it's only natural that we equip the companies that brought us *Police Academy 11*, Windows Vista, Milli Vanilli, and *Dancing With the Stars* with wire-cutters that allow them to disconnect anyone in the country on their own say-so, without proving a solitary act of wrongdoing.

But if that magic wire is indeed so trivial, they won't mind if we hold *them* to the same standard, right? The sloppy, trigger-happy litigants who sue dead people and children, who accused a laser printer of downloading the new Indiana Jones movie, who say that proof of wrongdoing is too much to ask for—if these firms believe that being disconnected from the internet is such a trivial annoyance, they should be willing to put up with the same minor irritation at corporate HQ and the satellite offices, right?

For Whom the Net Tolls

Rupert Murdoch wants to remake the web as a toll booth, with him in the collector's seat, but the net won't shift to his will

Just what, exactly, is Rupert Murdoch *thinking?* First, he announces that all of News Corp's websites will erect paywalls like the one employed by the *Wall Street Journal* (however, Rupert managed to get the details of the *WSJ*'s wall wrong—no matter, he's a "big picture" guy). Then, he announced that Google and other search engines were "plagiarists" who "rip off" News Corp's content, and that once the paywalls are up (a date that keeps slipping farther into the future, almost as though the best IT people work for someone who's not Rupert "I Hate the Net" Murdoch!) he'll be blocking Google and the other "parasites" from his sites, making all of News Corp's properties invisible to search engines. Then, as a kind of loonie cherry atop a banana split with extra crazy sauce, Rupert announces that "fair use is illegal" and he'll be abolishing it shortly.

What is he thinking? We'll never know, of course, but I have a theory.

First, the business of blocking search engines. Rupert has got dealmaker's flu, a bug he acquired when he bought MySpace and sold the exclusive right to index it to Google. This had the temporary effect of making Rupert look like a technology genius, as Google's putative payout for this right made the MySpace deal instantly profitable, at least on paper; meanwhile, MySpace's star was in decline, thanks to competition from Facebook, Twitter, and a million me-too social networking tools.

It also put ideas into Rupert's head.

You can practically see the maths on the blackboard behind his eyelids: exclusive deals + paywalls = money.

I think that Rupert is betting that one of Google's badly trailing competitors can be coaxed into paying for the right to index all of News Corp's online stuff, if that right is exclusive. Rupert is thinking that a company such as Microsoft will be willing to pay to shore up its also-ran search tool, Bing, by buying the right to index the fraction of a fraction of a sliver of a crumb of the internet that News Corp owns.

They'll be able to advertise: "We have Rupert's pages and Google doesn't, so search with us!" (Actually, they'll have to advertise: "We have Rupert's pages and Google doesn't, except MySpace, which Google has.")

Or maybe not—MySpace is not delivering the traffic Rupert guaranteed Google in his little deal, and Google may bail if there's a likely sucker on the line.

Maybe the target isn't Microsoft. Maybe it's some gullible startup that's even now walking up and down Sand Hill Road, the heart of Venture Capital Country in Silicon Valley, showing off a PowerPoint deck whose entire message can be summarised as: "You give us a heptillion dollars, we'll do exclusive search deals with Rupert and the other media behemoths, and we'll freeze Google out." I'd be surprised if such a pitch sold, though. What's the liquidity event that would return some profit to the VC? It's not *going* to be an IPO (Initial Public

Offering), not in today's regulatory climate. It'd have to be an acquisition, and the two most likely targets would be Google and News Corp.

Now, what about fair use being illegal? At a guess, I'd say that some Cardinal Richelieu figure in News Corp's legal department may have been passing some whispers to Rupert about international copyright law. Specifically, about the Berne Convention—a centuries-old copyright accord that's been integrated into many other trade agreements, including the World Trade Organization (WTO), and its "three-step test" for whether a copyright exemption is legal.

Copyright exemptions are all the rights that copyright gives to the public, not to creators or publishers, and "three-steps" describes the principles that Berne signatory countries must look to when crafting their own copyright exemptions.

Those three steps limit copyright exemptions to:

1. certain special cases...
2. which do not conflict with a normal exploitation of the work; and...
3. do not unreasonably prejudice the legitimate interests of the rights holder.

Now, arguably, many countries' fair dealing or fair use rules don't meet these criteria (the U.S. rules on VCRs, book lending, cable TV, jukeboxes, radio plays, and a hundred other cases are favourite villains in these discussions; but many European rules are also difficult to cram into the three-steps frame). And I've certainly heard many corporate law mover-shakers announce that, with the right lawsuit, you could get trade courts to force this country or that country to get rid of its fair dealing or fair use provisions.

However, this view of international copyright lacks an appreciation of the subtleties of international trade, namely: big, powerful countries can ignore trade courts and treaty rules when it's in their interest to do so, because no one can afford to stop trading with them.

The U.S. gets $1 trillion added to its GDP every year thanks to liberal fair use rules. If the WTO says that it has to ban video recorders or eliminate compulsory licenses on music compositions (or shut down search engines!), it will just ignore the WTO. The U.S. is an old hand at ignoring the United Nations. The U.S. owes billions to the UN in back-dues and shows no signs of repaying it. The fact that the WTO looks upon the U.S. with disapproval will cause precisely nothing to happen in the American legislative branch.

And, if the WTO tries to get other countries to embargo the U.S., it will quickly learn that China and other factory states can't afford to stop shipping plastic gewgaws, pocket-sized electronics, and cheap textiles to the United States.

And furthermore, other countries can't afford to boycott China—because those countries can't afford to allow a plastic gewgaw and cheap textile gap to emerge with America.

Of course, the elimination of fair use would present many problems to News Corp—because, as with all media companies, News Corp relies heavily on copyright exemptions to produce its own programming. I'm sure that, if there's a lawyer who put this idea into Rupert's head, she knows this. But I likewise believe that she would be perfectly willing to expand the legal department to the thousands of lawyers it would take to negotiate permission for all those uses if fair use goes away.

That's my theory: Rupert isn't a technophobic loon who will send his media empire to the bottom of the ocean while waging war on search engines. Instead, he's an out-of-touch moustache-twirler who's set his sights on remaking the web as a toll booth (with him in the collector's seat), and his plan hinges on a touchingly naive approach to geopolitics.

Either way, old Rupert shows signs of degenerating into a colourful Howard Hughes figure in a housecoat, demanding that reality shift to his will.

How Do You Know If Copyright Is Working?

A recurring question in discussions of digital copyright is how creators and their investors (that is, labels, movie studios, publishers, etc.) will earn a living in the digital era. But though I've had that question posed to me thousands of times, no one has ever said *which* creators and *which* investors are to earn a living, and what constitutes "a living." Copyright is in tremendous flux at the moment; governments all over the world are considering what their copyright systems should look like in the 21st century, and it's probably a good idea to nail down what we want copyright to *do.* Otherwise, the question, "Is copyright working?" becomes as meaningless as "How long is a piece of string?"

Let's start by saying that there is only one regulation that would provide everyone who wants to be an artist with a middle-class income. It's a very simple rule:

"If you call yourself an artist, the government will pay you £40,000 a year until you stop calling yourself an artist."

Short of this wildly unlikely regulation, full employment in the arts is a beautiful and improbable dream. Certainly, no copyright system can attain this.

If copyright is to have winners and losers, then let's start talking about who we want to see winning, and what victory should be.

In my world, copyright's purpose is to encourage the widest participation in culture that we can manage—that is, it should be a system that encourages the most diverse set of creators, creating the most diverse set of works, to reach the most diverse audiences as is practical.

That is, I don't want a copyright system that precludes making money on art, since there are some people who make good art who, credibly, would make less of it if there wasn't any money to be had. But at the same time, I don't think that you can judge a copyright system by how much money it delivers to creators—imagine a copyright system for films that allowed only one single 15-minute short film to be made every year, which, by dint of its rarity, turned over £1B. If only one person gets to make one movie, I don't care how much money the system brings in, it's not as good as one in which lots of people get to make lots of movies.

Diversity of participation matters because participation in the arts is a form of expression, and here in the West's liberal democracies, we take it as read that the state should limit expression as little as possible and encourage it as much as is possible. It seems silly to have to say this, but it's worth noting here, because when we talk about copyright, we're not just talking about who pays how much to get access to which art: we're talking about a regulation that has the power to midwife—or strangle—enormous amounts of expressive speech.

Here's something else copyright can't and won't do and doesn't do: deliver a market where creators (or investors) set a price for creative works, and audiences buy those works or don't, letting the best float to the top in a pure and free marketplace. Copyright has never really worked like this, and it certainly doesn't work like this today: for example, it's been more than a century since legal systems around the world took away songwriters' ability to control who performed their songs. This began with the first records, which were viewed as a form of theft by the composers of the day. You see, composers back then were in the sheet-music business: they used a copying device (the printing press) to generate a product that musicians could buy.

When recording technology came along, musicians began to play the tunes on the sheet music they'd bought into microphones and release commercial recordings of their performances. The composers fumed that this was piracy of their music, but the performers said, basically, "You sold us this sheet music—now you're telling us we're not allowed to play it? What did you think we were going to do with it?"

The law's answer to this was a Solomonic divide-the-baby solution: performers were free to record any composition that had been published, but they had to pay a set rate for every recording they sold. This rate was paid to a collective rights society, and today, these societies thrive, collecting fees for all sorts of "performances" where musicians and composers get little or no say—for example, radio stations, shopping malls, and even hairdressers buy licenses that allow them to play whatever music they can find. The music is sampled by more-or-less accurate means and dispersed to artists by more-or-less fair means.

Of course, some artists argue that the sampling and dispersal are unfair, but it's a rare artist who says that the principal of collective licensing is itself a form of theft.

No one wants to get a phone call every 15 minutes from some suburban barman who wants to know if playing their 20-year-old hit on the karaoke machine is going to cost 15p or 25p in license fees.

There is an ancient copyright agreement that Victor Hugo came up with called the Berne Convention that most western nations are parties to. If you read the agreement closely, it seems to make this whole business of blanket licensing illegal. When I've asked international copyright specialists how all these Berne nations can have radio stations and karaoke bars and hairdressers and such playing music without negotiating all their playlists one at a time, the usual answer is, "Well, technically, I suppose, they shouldn't. But there's an awful lot of money changing hands, mostly in the direction of labels and artists, so who's going to complain, really?"

Which is by way of affirming that grand old Americanism: Money talks and bullshit walks. Where the stiff-necked moral right of a copyright holder to control usage rubs up against the practicalities of allowing an entire industry's capacity for cultural exchange and use, the law usually responds by converting the moral right to an economic right. Rather than having the right to specify who may use your works, you merely get the right to get paid when the use takes place.

Now, on hearing this, you might be thinking, "Good God, that's practically Stalinist! Why can't a poor creator have the right to choose who can use her works?" Well, the reason is that creators (and, notably, their industrial investors) are notoriously resistant to new media. The composers damned the record companies for pirates; the record labels damned the radio for *its* piracy; broadcasters vilified the cable companies for taking their signals; cable companies fought the VCR for its recording "theft." Big Entertainment tried to kill FM radio, TV remote controls (which made it easy to switch away from adverts), jukeboxes, and so on, all the way back to the Protestant Reformation's fight over who got to read the Bible.

Given that new media typically allow new creators to create new forms of material that is pleasing to new audiences, it's hard to justify giving the current lotto-winners a veto over the next generation of disruptive technologies. Especially when the winners of today were the pirates of yesteryear. Turnabout is fair play.

So the best copyright isn't the one that lets every creator license every use of her work piecemeal. Instead, it's the system that allows for such licensing, except where other forms of licensing—or no licensing at all—makes sense. For example, in the USA, which has the largest, most profitable broadcast and cable industry in the world, the law gives no compensation rights to rights-holders for home recording of TV shows. There's no levy on blank cassettes or PVRs in exchange for the right to record off the telly. It's free, and it has conspicuously failed to destroy American TV.

There are whole classes of creation and copying that fall into this category: in fashion, for example, designs enjoy limited or no protection under the law. And each year's designer rags are instantaneously pirated by knock-down shops as soon as they appear on the runway. But should we protect fashion the way we do music or books? It's hard to see why, apart from a foolish consistency: certainly, every currently ascendant fashion designer who'd benefit from such a thing started out by knocking off other designers. And there's no indication that fashion is under-invested, or fails to attract new talent, or that there is a lack of new fashion available to the public. Creating exclusive rights for fashion designers might allow more money to be made by today's winners, but these winners are already making as many designs as they can, and so the net diversity of fashion available to the world would fall off.

Back to the question: what's a good copyright look like?

Well, it's got to be both evidence-based and balanced. For example, if architects come forward with the claim that they need to be able to control photos of their buildings or no one will invest in an architect's education, they'd better have some pretty compelling evidence to back up that claim. On the one hand, we have the incontrovertible fact that today, prospective architects spend a lot of money on professional training without any such guarantee.

Of course, it's easy to imagine that *more* people would enroll in architecture schools if designing a building gave you a copyright in its likeness—everyone who wanted to photograph a public road would have to pay you a license fee for the use of "your" building. But given that there's no evidence that architecture programmes are wasting away for want of students, and given that architects seem to be thriving as a trade everywhere, the evidence suggests that we don't need to give architects these rights.

That's evidence, but what about balance? Well, say

that tomorrow, the number of architects did shelve off radically, and no one could find anyone to draw up plans for a new conservatory or mansard roof anymore. How could we save architecture? Well, we *could* give architects a copyright in the likeness of their buildings, and essentially put architects in the rent-collecting business: rather than devoting all their time to designing buildings, architects would spend most of their time sending legal threats to sites like Flickr and Picasa and TwitPic whenever some poor sod uploaded a picture of his flat's exterior Christmas decorations and inadvertently violated the architect's copyright.

This would certainly make more money for some architects (especially ones whose buildings were situated near public webcams—everyone who operated one of those would have to stump up for a license!). But the public cost would be *enormous.* Instead of the mere absurdity of coppers going around ticking off tourists for photographing public buildings (as though bombing was a precision undertaking, requiring that terrorists photograph buildings in detail before wandering into them with bombs under their coats and blowing themselves up); we'd have vast armies of private security guards representing the far-flung descendants of Christopher Wren and that miserable bastard who designed the awful tower-block at the end of my road in 1965 or so, hassling anyone who took out a camera to snap a picture of the car that just ran them over, or their kids adorably eating ice cream, or their mates heaving up a kebab into the gutter after a night's revels.

Google Streetview would be impossible. So would

holiday snaps. Amateur photography. Fashion shoots. News photography. Documentary film-making. Essentially, the cost of recording your life as you live it, capturing your memorable moments, would go to infinity, as you had to figure out how to contact and buy licenses from thousands of obscure architects or their licensees. Surely in this case, the costs outweigh the benefits (and yes, I'm perfectly aware that certain European countries were stupid enough to give architects this right—there are also places in the world that prohibit women from driving cars, where they chop down rainforests to graze cattle, and where the used car adverts feature florid men wearing foam cowboy hats screaming into a camera—if everyone in France jumped off the Eiffel Tower, would you do it too?).

So a balanced and evidence-based copyright policy is one that requires creators to show a need for protection, and also that the protection sought will deliver more benefit than the cost it implies.

How would this apply to the internet? Take music downloads: by the music industry's own account, the pay-per-download systems only capture a minute fraction of the music traded on the net. But a blanket license that ISPs could opt into that entitled the ISP's customers to download and share all the music they wanted would deliver evergreen profits to the record industry—without necessitating spying, lawsuits, and threats of disconnection from the internet. If the price was right, practically every ISP would opt into the system, since the cost of the legal headaches attending the operation of a service without such a license would be more expensive than getting legit. Then we could focus on making the collection and dispersal of fees and the sampling of music downloading as transparent as possible, bringing 21st-century metrics to bear on making sure that artists are fairly compensated (rather than spending vast sums figuring out which music fans to send legal threats to this month).

Now, take $300M CGI summer blockbuster films: if the producers of these things are to be believed, the ongoing capacity to produce glitzy, big budget productions demands that services like YouTube be shut off (see, for example, Viacom's lawsuit against Google over YouTube). If this is true—I'm no movie exec, maybe it is—then we need to ask ourselves the "balance" question: YouTube's users produce 29 hours of video every *minute* and the vast majority of it is not infringing TV and movie clips, it is independently produced material that accounts for more viewer-minutes than television. So, the big studios' demand amounts to this: "You must shut down the system that delivers billions of hours of enjoyment to hundreds of millions of people so that we can go on delivering about 20 hours' worth of big budget film every summer."

To me, this is a no brainer. I mean, I love sitting in an air-conditioned cave watching Bruce Willis beat up a fighter jet with his bare hands as much as the next guy, but if I have to choose between that and *all of YouTube,* well, sorry Bruce.

The rejoinder I hear from the film industry in these discussions is downright bizarre: they cite the fact that all those billions of hours' worth of material on YouTube cost very little to make, and consequently, YouTube is able to pay very small sums of money in ad revenue and still get all that video. To hear an industrialist damning a competitor because he's figured out a way of making a competing product that costs a lot less is just weird. There is no virtue in spending a lot of money. Anyone can do it. Spending small sums of money to make something great—well, that's just magic.

News Corp Kremlinology: What Do the *Times* Paywall Numbers Mean?

This week, Rupert Murdoch's News Corp International released a set of pubic figures on the performance of the paywall it put around the *Times* last July. The takeaway from the press release was rosy: The *Times* had about 200,000 paid users, 100,000 of whom were digital-only customers (meaning that the other 100,000 were print subscribers who'd gotten a free online sub bolted on to their existing offer). What's more, those 200,000 precious paid users were worth more to the *Times*, because the personal information they surrendered in the payment process could be used to better-target adverts to them, thus the *Times* could command a higher advertising rate in its paywalled incarnation.

Fundamentally, the question News Corp is trying to answer is, "Will the *Times* make more money with a paywall?" And the figures we've just seen do nothing to answer this question. Rather, the *Times* seems to think that the new figures prove something else: "People are willing to pay for their news." I don't think that anyone has ever disputed that someone, somewhere, was willing to pay for the *Times*, though: surely the important question, from a business perspective, is, "Will adding a paywall increase your profits?"

If these numbers were supposed to serve as validation for the paywall business model, they fell short of the mark. The coarseness of these figures is such that a multitude of business sins could be hidden within them.

To try to get to the bottom of this, I spoke to a News Corp spokesperson who—bizarrely—asked not to be identified by name (I've never encountered an anonymous official spokesperson before and I was pretty surprised at this request, especially as the figures the spokesperson gave me are "all in the public domain").

Here are the questions News Corp will need to answer if it wants to offer up the *Times* paywall as a success:

What sort of purchases are the paid subscribers making?

There are multiple retail offerings for the *Times*: you can buy a monthly subscription for £8.66, which includes iPad access (as well as access via other mobile device apps). You can buy a month's worth of iPad-only access for £9.99 (yes, the *Times* costs more as an iPad-only offering than it does if you get the iPad and the full web access together—go figure).

Then there are the lower-cost options. You can get a month's introductory offer to the *Times* for a mere £1. You can also get a day pass to the site for £1 (it costs the same to access the site for one day and one month—but the difference is that you don't have to remember to unsubscribe at the end of the day lest you be signed on for indefinite monthly £8.66 payments). You can also get free access to the *Times* with your TalkTalk mobile phone subscription.

The 100,000-odd paid users who pay extra for the *Times* are a mix of all these numbers, and News Corp won't disclose the nature of the mix, though the anonymous official spokesperson said that they do have these figures—which is good! If you're going to try something like this, you'd be mad not to audit the performance of all your offerings very closely.

Here's what the *Times* will say: about 50,000 of the current paid users are on a monthly subscription of some sort: £8.66, £1, or free with a TalkTalk subscription. They will not disclose how many £1 trial users turn into £8.66 users, or how many sustain their £8.66 subscription into the second or third month. However, the anonymous official spokesperson did say that whichever users are remaining after three months are more than 90 percent likely to stump up for a fourth month. From this, I think we can safely assume that lots *less* than 90 percent of paid users stick around for a second month, and of those, less than 90 percent sustain themselves for a fourth month.

But the *Times* isn't saying.

The remaining 50,000, of course, are people who paid £1 for a single day's access. Some number of these converted to monthly subscribers. Some number bought a second article. How many? The *Times* isn't saying.

So: best case: there are 50,000 paid subscribers, all of whom got there by paying £1 for an article, converted immediately to £1 monthly subscriptions, and now pay £8.66 every month (or £9.99 in the case of iPad users who want to pay extra for the privilege of not being allowed to access the website).

Worst case: 50,000 people tried a day-pass and buggered off. 20,000 TalkTalk subscribers got a free subscription with their phone which they may or may not know or care about. 5,000 people use it with an iPad. 75,000 people tried a £1 month trial. 40,000 of them signed up for a second month, 30,000 of them for a third, and 25,000 stayed on for a fourth month.

I don't know which one is closer to the truth, because the *Times* isn't saying. But I do know that when there was a positive number—more than 90 percent renewal at the third month—the figure was readily available, which leaves the distinct impression that all the undisclosed numbers are less than stellar.

How much do advertisers value the additional information the *Times* can supply about paying users?

The anonymous official spokesperson told me that about 50 percent of the *Times*'s bottom line comes from advertising, and that the number of unique users visiting the site has fallen from about 20,000,000 per month to 200,000 at present—a drop-off of about 99 percent (and only half of those are paying separately for online access, which means than less than one half of one percent of the *Times* readership has been willing to spend £1 or more to access the site).

The *Times* is betting that this drop-off can be overcome with higher advertising rates. So how much more are advertisers willing to spend to reach these logged in users?

The *Times* isn't saying.

While the *Times*'s print edition has a published rate-card (as do many of News Corp's newspapers' online editions, such as the *Wall Street Journal*, the *Sun* and *News of the World*), its online edition rate-card is confidential (though it wasn't, prior to the paywall). So there's no way to know how much the *Times* is asking advertisers to pay for placement on the paywalled site.

What's more, the *Times* has opted out of the national, industry-standard circulation audits, making the whole venture into more of a black box. The anonymous spokesperson wouldn't rule out opting back into independent circulation audits, but made no promises either.

What does it cost to get a subscriber?

The *Times*'s paywall was attended by an enormous amount of (justifiable) publicity as it was in itself a newsworthy event. But this free publicity was augmented with an enormous marketing blitz in print, billboard, TV, etc.—a campaign that brought in 100,000 customers. How much did this campaign cost? The *Times* isn't saying.

This is important. A well-executed and well-financed advertising campaign can get a couple hundred thousand people to try *anything*—give me £5,000 to spend reaching every person in Britain and I'll find you 200,000 people who'll spend a pound to rub blue mud in their navels on a trial basis. To be profitable, your marketing costs have to be lower than the income they generate.

Finally, there are some miscellaneous questions for which it'd be nice to have answers. For example: the *Times* gave free subscriptions to 150,000 of its print subscribers, About 100,000 of those subscribers tried the freebie out. How active are those two-thirds who took the plunge? Do they come back daily? Weekly? Monthly? The *Times*'s anonymous spokesperson said that they were "very active" but wouldn't say how many had logged in in the past 30 days.

So, what are we meant to make of the *Times*'s latest numbers? Well, perhaps the answers to the questions above are extremely flattering to the *Times* and its digital strategists, and they're withholding them (out of modesty? in order to make a big splash later?). On the other hand, perhaps the *Times* has spent an enormous amount of money on a plan that chased off 99 percent of its readers, and money hasn't yet rushed in to fill the vacuum they left behind.

Only the *Times* knows, and they're not saying.

Persistence Pays Parasites

My friend Katherine Myronuk once told me, "All complex ecosystems have parasites." She was talking about spam and malware (these days they're often the same thing) and other undesirable critters on the net. It's one of the smartest things anyone's ever said to me about the net—and about the world. If there's a niche, a parasite will fill it. There's a reason the cells of the organisms that live in your body outnumber your own by 100 to one. And every complex system has unfilled niches. The only way to eliminate unfilled niches is to keep everything simple to the point of insignificance.

But even armed with this intelligence, I've been pretty cavalier about my exposure to net-based security risks. I run an up-to-date version of a very robust flavor of GNU/Linux called Ubuntu, which has a single, easy-to-use interface for keeping all my apps patched with the latest fixes. My browser, Firefox, is far less prone to serious security vulnerabilities than dogs like Internet Explorer. I use good security technology: my hard drive and backup are encrypted, I surf through IPREDator (a great and secure anonymizer based in Sweden), and I use GRC's password generator to create new, strong passwords for every site I visit (I keep these passwords in a text file that is separately encrypted).

And I'm media-literate: I have a good nose for scams and linkbait, I know that no one's planning to give me millions for aiding in a baroque scheme to smuggle cash out of Nigeria, and I can spot a phishing email at a thousand paces.

I know that phishing—using clever fakes to trick the unsuspecting into revealing their passwords—is a real problem, with real victims. But I just assumed that phishing was someone *else's* problem.

Or so I thought, until I got phished last week.

Here's the thing: I thought that phishers set their sights on a certain kind of naive person, someone who hadn't heard all the warnings, hadn't learned to be wary of their attacks. I thought that the reason that phishers sent out millions of IMs and emails and other messages was to find those naifs and ensnare them.

But I'm not one of those naifs. I'd never been tricked, even for a second, by one of those phishing messages.

Here's how I got fooled. On Monday, I unlocked my Nexus One phone, installing a new and more powerful version of the Android operating system that allowed me to do some neat tricks, like using the phone as a wireless modem on my laptop. In the process of reinstallation, I deleted all my stored passwords from the phone. I also had a couple of editorials come out that day, and did a couple of interviews, and generally emitted a pretty fair whack of information.

The next day, Tuesday, we were ten minutes late getting out of the house. My wife and I dropped my daughter off at the daycare, then hurried to our regular coffee shop to get take-outs before parting ways to go to our respective offices. Because we were a little late arriving, the line was longer than usual. My wife went off to read the free newspapers, I stood in the line. Bored, I opened up my phone fired up my freshly reinstalled Twitter client and saw that I had a direct message from an old friend in Seattle, someone I know through fandom. The message read "Is this you????" and was followed by one of those ubiquitous shortened URLs that consist of a domain and a short code, like this: http://owl.ly/iuefuew.

I opened the link with my phone and found that I'd been redirected to the Twitter login page, which was prompting me for my password. Seeing the page's URL (truncated in the little phone-browser's location bar as "http://twitter....") and having grown accustomed to re-entering all my passwords since I'd reinstalled my phone's OS the day before, I carefully tapped in my password, clicked the login button, and then felt my stomach do a slow flip-flop as I saw the URL that my browser was contacting with the login info: http://twitter.scamsite.com (it wasn't really scamsite, it was some other domain that had been hijacked by the phishers).

And that's when I realized that I'd been phished. And it was bad. Because I'd signed up for Twitter *years* ago, when Ev Williams, Twitter's co-founder sent me an invite to the initial beta. I'd used a password that I used for all kinds of sites, back before I started strictly using long, random strings that I couldn't remember for passwords. In defense of the old me, I only used that password for unimportant sites, like services that friends wanted me to sample in beta.

But unimportant sites have a way of becoming important. I've got 40,000+ Twitter followers, and if my account was hijacked, the hijackers could do great damage to my reputation and career through their identity theft.

What's more, Twitter isn't the only place where I used my "low-security" password that has turned into a high-security context, which means that hijackers could conceivably break into lots of interesting places with that information.

So I sat down at a table, kissed my wife goodbye, got my laptop out, and started changing passwords all over the net. It took hours (but at least I've expunged that old password from my existing accounts, I think). By the time I finished, three more copies of the phishing scam had landed in my Twitter inbox. If they'd come a few minutes earlier, the multiple copies would have tripped my radar and I would have seen them for a scam. The long process gave me lots of time to reconsider my internal model of how phishing works.

Phishing isn't (just) about finding a person who is technically naive. It's about attacking the seemingly impregnable defenses of the technically sophisticated until you find a single, incredibly unlikely, short-lived crack in the wall.

If I hadn't reinstalled my phone's OS the day before. If I hadn't been late to the cafe. If I hadn't been primed to hear from old friends wondering if some press mention was me, having just published a lot of new work. If I hadn't been using a browser that didn't fully expose URLs. If I hadn't used the same password for Twitter as I use for lots of other services. If I'd been ten minutes *later* to the cafe, late enough to get multiple copies of the scam at once—for the want of a nail, and so on.

But all the stars aligned for that one moment, and in that exact and precise moment of vulnerability, I was attacked by a phisher. This is eerily biological, this idea of parasites trying every conceivable variation, at all times, on every front, seeking a way to colonize a host organism. The net's complex ecosystem is so crowded with parasites now that it is a sure bet that there will be a parasite lurking in the next vulnerable moment I experience, and the next. And I will have vulnerable moments. We all do.

I don't have a solution, but at least I have a better understanding of the problem. Falling victim to a scam isn't just a matter of not being wise to the ways of the world: it's a matter of being caught out in a moment of distraction and of unlikely circumstance.

Like Teenagers, Computers Are Built to Hook Up

Demanding that users abstain from net will never work when they need it for their jobs. Better to practise safe hex

Real-world disease-prevention often means checking in the word "should" at the door. Take abstinence programmes: whether or not you think kids *should* be having sex, you can't miss the fact that they *are* having sex. If you want kids to stay disease-free and healthy, you have to provide them with the tools and skills to have sex while doing so. The facts speak for themselves; countries where abstinence is the primary mitigation strategy have higher rates of teenage pregnancy and sexually transmitted infections than countries where sexual education and free birth control and condoms are the rule.

Of course, it's worth asking why kids are having sex and whether you can do something about that fact, too. The researcher danah boyd has identified at-risk kids haunting sexually explicit chatrooms—and it's there that predators go to find prey, not random messageboards or chatrooms (boyd likens the idea that predators will find victims on random MySpace pages to the idea that they would pick phone numbers at random and dial them). If you want to make kids *really* safe, it's worth looking into the factors that send kids out looking for trouble.

There's a lesson for IT security in here.

Every time a state secret disappears from an internet-connected PC, every time a hospital computer reboots itself in the middle of a surgical procedure because it has just downloaded the latest patch, every time an MRI machine gets infected with an internet worm, I hear security experts declaiming, "Those computers should never be connected to the internet!" and shaking their heads at the foolish users and the foolish IT department that gave rise to a situation where sensitive functions were being executed on a computer connected to the seething, malware-haunted public internet.

But no amount of head-shaking is going to change the fact that computers, by and large, get connected. It's what they're designed to do. You might connect to the internet without even meaning to (for example, if your computer knows that it's allowed to connect to a BT Wi-Fi access point, it will connect and disconnect from hundreds of them if you carry it with you through the streets of London).

Operating systems are getting more promiscuous about net connections, not less: expect operating systems to start seeking out Bluetooth-enabled 3G phones and using them to reach out to the net when nothing else is available.

All evidence suggests that keeping computers off the internet is a losing battle. And even if you think you can discipline your workers into staying offline, wouldn't it be lovely if you had a security solution that worked *even if* someone broke the rules? "You shouldn't be having net at your age, but if you do, you should at least practice safe hex."

A good security system—especially for sensitive

machines—should contemplate the possibility that a computer is going to be connected to the net even if that's not supposed to happen: needless services turned off, appropriate firewall rules (including rules that distrust the LAN as well as the WAN), good auto-update policies that require human intervention.

But IT departments need to go beyond defense in depth. To effectively secure a network, you need to become an epidemiologist of your users' unsafe activity. Did the radiologist plug the ethernet into the MRI machine because she needed to update the controller software with a new version in order to get her job done? Are the operating theatre's machines on the LAN because surgeons have followed the entire rest of the world in outsourcing their remembrance of petty facts to search engines? Does that defence contractor carry his sensitive materials on his laptop because he is collaborating with hundreds of other contractors in a huge, complex endeavour only possible with networked communications?

Users will always prioritise getting their job done over honouring your network policy, and who can blame them? If network policy breaches aren't followed up with safe solutions to users' demonstrated needs, they'll keep on happening, no matter how much security you put between your users and their duties.

In the era of cheap and easy virtualisation and sandboxing, there's no reason users shouldn't be able to partition their computers into "dirty" public-facing sides and "clean" private sides. Of course, a user might subvert this separation deliberately, but the only way to comprehensively prevent that from occurring is to make it possible for a user to get the job done without needing to do so.

Just like the parents who are running around shagging their brains out while preaching abstinence, IT departments are generally happy to step outside the boundaries they set out for their users in order to get *their* jobs done. Teenagers aren't the only people who ignore abstinence programs—users and kids can sniff out hypocrisy a mile away.

Promoting Statistical Literacy: A Modest Proposal

Why do our institutions—particularly banks—fail to grasp the most rudimentary basics of password security?

Here's a modest proposal: what if the government took it on board to promote a reasonable, sane grasp of risk, security, and probability? Or, if you're a "Big Society/Small Government" LibCon, how about a more modest mandate still: we could ask the state to leave off *promoting* statistical innumeracy and the inability to understand risk and reward.

Start with the lottery: in the U.S., its slogan is "Lotto: You've Got to Be in It to Win It." A more numerate slogan would be "Lotto: Your Chance of Finding the Winning Ticket in the Road Is Approximately the Same as Your Chance of Buying It." The more we tell people that there is a meaning gap between the one-in-a-squillion chance of finding the winning ticket and the one-in-several-million chance of buying it, the more we encourage the statistical fallacy that events are inherently more likely if they're very splashy and interesting to consider.

This is the same reasoning that causes parents to run in circles squawking in terror at the thought of paedophiles stalking their kiddies, even as they let Junior ride in the car without his seatbelt—auto fatalities being orders of magnitude more common than random paedophile attacks. (Of course, the most likely paedophile in your child's life is you or your spouse, or a close friend, relative, or authority figure.) Preparing for the unlikely while neglecting the (relatively) common is a terrible way to make the world safer for you and yours.

Banish the lotto? Wouldn't that mean losing all the lovely money extracted by way of a voluntary tax on innumeracy? Perhaps, but if getting rid of the lottery could give rise to a modest increase in common sense about risk and security, think of the society-wide savings in money not spent on alarmist newspapers, quack child-protection schemes, MMR scares, and the like!

Once we get rid of the lottery, let's attack the banks. It's not bad enough that they collect enormous bonuses at public expense while destroying the economy; they also systematically disorder our capacity to understand risk and security through an ever-more-farcical stream of "compliance" hoops and bizarro-world "security" measures!

For example, my own bank, the Co-op, recently updated its business banking site (the old one was "best viewed with Windows 2000!"), "modernising" it with a new two-factor authentication scheme in the form of a little numeric keypad gadget you carry around with you. When you want to see your balance, you key a Pin into the gadget, and it returns a *10-digit* number, which you then have to key into a browser-field that helpfully masks your keystrokes as you enter this gigantic one-time password.

Don't get me wrong: two-factor authentication makes perfect sense, and there's nothing wrong with using it to keep users' passwords out of the hands of keyloggers and other surveillance creeps. But a system that locks users out after three bad tries does not need to generate a 10-digit one-time password: the likelihood of guessing a modest four- or five-digit password in three tries is small enough that no appreciable benefit comes out of the other digits (but the hassle to the Co-op's many customers of these extra numbers, multiplied by every login attempt for years and years to come, is indeed appreciable).

As if to underscore the Co-op's security illiteracy, we have this business of masking the one-time Pin as you type it. The whole *point* of a one-time password is that it *doesn't matter* if it leaks, since it only works *once*. That's why we call it a "one-time Pin." Asking customers to key in a meaningless 10-digit code perfectly, every time, without visual feedback, isn't security. It's sadism.

It gets worse: the Pin you use with the gadget is your basic four-digit Pin, but numbers can't be sequential. This has the effect of reducing the keyspace by an enormous factor—a bizarrely contrarian move from a bank that "improves" its security by turning this constrained four-digit number into a whopping 10-digit one. Does the Co-op love or loathe large keyspaces? Both, it seems.

It's not just the Co-op, of course—this is endemic to the whole industry. For example, Citibank UK requires you to input your password by chasing a tiny, on-screen, all-caps password with your mouse-pointer, in the name of preventing a keylogger from capturing your password as you type it. This has the neat triple-play effect of slicing the keyspace in half (and more) by eliminating special characters and lower-case letters; incentivising customers to use shorter, less secure passwords because of the hassle of inputting them; *and* leaving the whole thing vulnerable to screen-loggers that simply make movies of which keys you mouse over.

But I quit Citibank, and I still use the Co-op for my commercial banking out of some bloody-minded, bolshy commitment to "good" banking, even though they require that foreign drafts be requested by means of faxes on headed paper (neither faxes nor headed paper being any sort of security system) and so on. Possibly it's because they occasionally see reason, as when I opened an account with my wife and discovered that I could either bring certified copies of both our passports to a branch; or I could bring my wife and her passport to a branch. The fact that my wife didn't have to be present in order to get a certified copy was a difficult concept for the Co-op to master, but once it did, a compliance officer agreed that this meant I should be able to simply show up at a branch with both passports without throwing money at some rich solicitor for the privilege of getting his stamp at the bottom of a photocopy.

It wasn't easy—the branch staff couldn't believe that I had won an exception to this weird policy—but in the end, they opened the account for me. Now, like a mouse that's found an experimental lever that only *sometimes* gives up a pellet, I find myself repeatedly pressing it, hoping to hit on the magical combination that will get my bank to behave as though security was something that a reasonable, sane person could understand, as opposed to a magic property that arises spontaneously in the presence of sufficient obfuscation and bureaucracy.

The great irony, of course, is that all the banks will tell you that they're only putting you through the

Hell of Nonsensical Security because the FSA or some other authority have put them up to it. The regulators strenuously deny this, saying that they only specify principles—"thou shalt know thy customer"—not specific practices.

Which brings me back to my modest proposal: let's empower our regulators to fine banks that create nonsensical, incoherent security practices involving idolatrous worship of easy-to-forge utility bills and headed paper, in the name of preserving our national capacity to think critically about security.

Even if it doesn't kill the power of the tabloids to sell with screaming headlines about paedos, terrorists, and vaccinations, it would, at least, be incredibly satisfying to keep your money in an institution that appears to have the most rudimentary grasp of what security is and where it comes from.

Personal Data Is as Hot as Nuclear Waste

We should treat personal electronic data with the same care and respect as weapons-grade plutonium—it is dangerous, long-lasting and once it has leaked there's no getting it back

When HM Revenue & Customs haemorrhaged the personal and financial information of 25 million British families in November, wags dubbed it the "Privacy Chernobyl," a meltdown of global, epic proportions.

The metaphor is apt: the data collected by corporations and governmental agencies is positively radioactive in its tenacity and longevity. Nuclear accidents leave us wondering just how we're going to warn our descendants away from the resulting wasteland for the next 750,000 years while the radioisotopes decay away. Privacy meltdowns raise a similarly long-lived spectre: will the leaked HMRC data ever actually vanish?

The financial data in question came on two CDs. If you're into downloading movies, this is about the same size as the last couple of Bond movies. That's an incredibly small amount of data—my new phone holds 10 times as much. My camera (six months older than the phone) can only fit four copies of the nation's financial data.

Our capacity to store, copy, and distribute information is ascending a curve that is screaming skyward, headed straight into infinity. This fact has not escaped the notice of the entertainment industry, where it has been greeted with savage apoplexy.

Wet Kleenex

But it seems to have entirely escaped the attention of those who regulate the gathering of personal information. The world's toughest privacy measures are as a wet Kleenex against the merciless onslaught of data acquisition. Data is acquired at all times, everywhere.

For example, you now *must* buy an Oyster Card if you wish to buy a monthly travelcard for London Underground, and you are *required* to complete a form giving your name, home address, phone number, email, and so on in order to do so. This means that Transport for London is amassing a radioactive mountain of data plutonium, personal information whose limited value is far outstripped by the potential risks from retaining it.

Hidden in that toxic pile are a million seams waiting to burst: a woman secretly visits a fertility clinic, a man secretly visits an HIV support group, a boy passes through the turnstiles every day at the same time as a girl whom his parents have forbidden him to see; all that and more.

All these people could potentially be identified, located, and contacted through the LU data. We may say we've nothing to hide, but all of us have private details we'd prefer not to see on the cover of tomorrow's paper.

How long does this information need to be kept private? A century is probably a good start, though if it's the kind of information that our immediate descendants would prefer to be kept secret, 150 years is more like it.

Call it two centuries, just to be on the safe side.

Weapons-grade data

If we are going to contain every heap of data plutonium for 200 years, that means that every single person who will ever be in a position to see, copy, handle, store, or manipulate that data will have to be vetted and trained every bit as carefully as the folks in the rubber suits down at the local fast-breeder reactor.

Every gram—sorry, byte—of personal information these feckless data-packrats collect on us should be as carefully accounted for as our weapons-grade radioisotopes, because once the seals have cracked, there is no going back. Once the local sandwich shop's CCTV has been violated, once the HMRC has dumped another 25 million records, once London Underground has hiccoughed up a month's worth of travelcard data, there will be no containing it.

And what's worse is that we, as a society, are asked to shoulder the cost of the long-term care of business and government's personal data stockpiles. When a database melts down, we absorb the crime, the personal misery, the chaos, and terror.

The best answer is to make businesses and governments responsible for the total cost of their data collection. Today, the PC you buy comes with a surcharge meant to cover the disposal of the e-waste it will become. Tomorrow, perhaps the £200 CCTV you buy will have an added £75 surcharge to pay for the cost of regulating what you do with the footage you take of the public.

We have to do something. A country where every snoop has a plutonium refinery in his garden shed is a country in serious trouble.

Memento Mori

I'm often puzzled by how *satisfying* older technology is. What a treat it is to muscle around an ancient teletype, feeding it new-old paper-tape or rolls of industrial paper with the weight of a bygone era. What pleasure I take from the length of piano roll I've hung like a banner from a high place in every office I've had since 2000. How much satisfaction I derive from the racing works of the 1965 mechanical watch I received for a Father's Day present this year, audible in rare moments of ambient silence or when my hand strays near my ear, going tick-tick-tick-tick like the pattering heart of a pet mouse held loosely in my hand. What joy I take in the 19th-century Chinese copy of a 14th-century French skeleton clock I gave my wife as a wedding present, tick-tocking loudly in proud brassworks under the tall bell jar that protects its delicate exposed mechanisms!

The standard explanation for the attractiveness of this old stuff is simply that They Made It Better In The Old Days. But this isn't necessarily or even usually true. Some of my favorite old technologies are as poorly made as the most throwaway products of the sweatshops of today's Pearl River Delta. Take that piano roll, for example: a cheap and flimsy entertainment, hardly made to be appreciated as an artifact in itself. And those rattling machine-gun teletypes and caterpillar-feed daisy-wheel printers? Feeding them was a dark art, like sprocketing a filmstrip on the most clapped out projector in the school's AV closet—they have all the engineering elegance of a plastic cap gun that falls apart after the first roll of caps has run through it.

Today, I have a different answer. Sitting beside me as I type this is an open USB SATA enclosure with a 512GB Kingston solid state drive fitted to it, and the case's lights are strobing like the world's tiniest rave as my computer idly writes its partitions and filesystems. Every time I look at this thing, I giggle. I've been giggling all afternoon.

I got my first personal computer in 1979, an Apple II+, and it came with 48K of main memory (before we got a personal computer in the house, we used teletypes with acoustic couplers to talk to PDPs at the University of Toronto). I remember well the day we upgraded the RAM to 64K, my father slotting in the huge board reverently, knowing that it represented $495 worth of our family's tight technology budget (about $1,500 in today's money). What I really remember, of course, is the screaming performance boost that we got from that board.

That set a pattern for the rest of my computing life: RAM, RAM, RAM. The more memory you stuffed in the machine, the better it got. It was like Tabasco Sauce for computing: it improved everything you added it to. But unlike Tabasco, RAM cost the world.

I remember the first time RAM made me laugh.

It was the mid-1990s. I was learning to do Unix systems integration for prepress, and commuting back and forth to Silicon Valley. My mentor and friend, Miqe, would drive us from San Francisco to Cupertino every morning at oh-dark-hundred to beat the commuter traffic. We'd talk computers. We were doing prepress installations, going into shops where every designer had two or more brand-new Quadras on her desk. After completing a job, she'd hit cmd-P and the Quadra would start ripping, sometimes taking three or four days to complete the job. While that was going on, she'd move to the next Quadra and start working on the next job. One of the major computing bottlenecks was paging memory out to disk, and, of course, every machine already had as much RAM as it could handle (136MB).

Miqe and I got to talking about how much these machines would benefit from having lots of RAM. We talked about the performance improvements we'd be able to get with an unthinkable 500MB of RAM. Then we thought about 1GB of RAM and all we could do with it. Finally, we strained our imaginations out to their outer limits and tried to imagine computing at one terabyte of RAM.

And we started to laugh. First a little, then a lot. This substance that cost more than its weight in gold, that solved all our problems—sometime in our lifetimes it would be so cheap and abundant that we would have literally *unimaginable* amounts of it.

And that's why I've been giggling at this half-terabyte RAM (OK, RAM-like) drive that I just spent $1,500 on—the same sum Dad parted with for that 64K upgrade card 30 years ago.

Which brings me back to these beautiful old machines I've got around my office, from the 300,000-year-old stone axe-head to the rusting, nonfunctional wind-up bank. I don't have these here because they're inherently well-made or beautiful. I have them here because they are uproarious, the best joke we have. They are the continuous, ever-delightful reminder that we inhabit a future that rushes past us so loudly we can barely hear the ticking of our watches as they are subsumed into our phones, which are subsumed into our PCs, which are presently doing their damndest to burrow under our skin.

The poets of yore kept human skulls on their desks as mementos mori, reminders of humanity's fragility. I keep these old fossil machines around for the opposite reason: to remind me, again and again, of the vertiginous hilarity of our age of wonders.

Love the Machine, Hate the Factory

We've heard a lot about how scary the industrial revolution was—the dislocations it wrought on the agrarian population of the early 19th century were wrenching and terrible, and the revolution was a bloody one. From that time, we have the word "Luddite," referring to uprisings against the machines that were undoing ancient ways of living and working.

But the troubles of the 1810s were only the beginning. By the end of the century, the workplace was changing again. Workers who'd adapted over three generations to working in factories at machines, rather than tilling the land and working in small cottage workshops, once again found their lives being dramatically remade by the forces of capital, through a process called "scientific management."

Scientific management (which was also called "Taylorism" or "Scientific Taylorism," for its most prominent advocate, Frederick Winslow Taylor) was built around the idea of reducing a manufacturing process to a series of optimized simple steps, creating an assembly line where workers were just part of the machine. Each worker's movements were as scripted as those of a cog or piston, defined by outside observers who sought to make the work go as smoothly as possible, with as few interruptions as possible.

Taylor, Henry Ford, and Frank and Lillian Gilbreth used time-motion studies, written logbooks, high-speed photography, and other empirical techniques to find wasted motions, wasted times, potential logjams in the manufacturing process, and practically every industry saw massive increases in productivity thanks to their work. The Gilbreths' research gave us modern surgical procedure, touch-typing, and a host of other advances to human endeavour.

But all this gain was not without cost. The "unscientific" worker personally worked on several tricky stages of manufacture, often seeing a project through from raw materials to finished product. He or she could choose how to sit, which tool to use when, and what order to complete the steps in. If it was a sunny day with a fine autumn breeze, the worker could choose to plane the joints and keep the smell of the leaves in the air, saving the lacquer for the next day. Workers who were having a bad day could take it easy without holding up a production line. On good days, the work could fly past without creating traffic jams further down the line.

For every gain in efficiency, scientific management exacted a cost in self-determination, personal dignity, and a worker's connection with what s/he produced.

For me, the biggest appeal to steampunk is that it exalts the machine and disparages the factory (this is the motto of the excellent and free *Steampunk* magazine: "Love the Machine, Hate the Factory"). It celebrates the elaborate inventions of the scientifically managed enterprise, but imagines those machines coming from individuals who are their own masters. Steampunk doesn't rail against efficiency—but it never puts efficiency ahead of self-determination. If you're going to raise your workbench to spare your back, that's *your* decision, not something imposed on you from the top down.

Here in the 21st century, this kind of manufacture finally seems in reach: a world of desktop fabbers, low-cost workshops, communities of helpful, like-minded makers puts utopia in our grasp. Finally, we'll be able to work like an artisan and produce like an assembly line.

Untouched by Human Hands

Worrying about the origin of a product is as old—at least—as the Florentine tradition of baking unsalted bread so as to avoid buying salt from the archrivals in neighboring Pisa. Whether it's Gandhi's defiant salt-making, a "Buy American" or "Look for the union label" campaign, a PETA anti-fur protest, or the "organic" badge on a carrot at the grocery store, the act of consumption has always been fraught with ethical conundra and considerations.

I spent most of 2008 researching my novel *For the Win*, which is largely set in the factory cities of South China's Pearl River Delta. If you own something stamped "MADE IN CHINA" (and you do!), chances are it was made in one of these cities, where tens of millions of young women have migrated since the combination of Deng Xiao Ping's economic reforms at the World Trade Organization agreement set in motion the largest migration in human history.

It's difficult to characterize the products of these factories: everything from high-priced designer goods to the cheapest knock-offs originate there (on average, one container per second leaves South China for America, every second of every hour of every day). But there is one characteristic almost all these products share: they are produced on an assembly line, and they are supposed to look like it. Very few of the goods in the stream of mainstream commerce are hand-made, and when they are, they are supposed to look like they aren't (the obvious exception is distressed clothing, but the impression of a unique wear patina is quickly dispelled when you see twenty identically patched and worn jean-jackets for sale at an outlet mall, in small, medium, and large).

Indeed, it's almost impossible to imagine a mainstream store that sold handmade goods for the purpose of daily use by average people. The notion of "hand-made" has undergone several revolutions in the past century, its meaning alternating between "precious and artisanal" to "cheap and inferior." Artisanal fashions have likewise swung between the two poles of "rough and idiosyncratic" and "all seams hidden, every rough edge sanded away."

Today, the fit and finish that the most careful, conscientious artisan brings to her creations usually ends up making it look machine-finished, injection-moulded—seamless because it was untouched by human hands, not because it was lovingly handled until every blemish was gone. What's more, the increasing awareness of the environmental and human cost of intensive manufacturing has started to give factory goods a whiff of blood and death. Your new mobile phone was made by a suicidal Foxconn worker, from coltan mud extracted by slaves in a brutal dictatorship, shipped across the ocean in a planet-warming diesel freighter, and is destined to spend a million years in a landfill, leeching poison into the water table.

Which leads me to wonder: is there a boardroom somewhere where a marketing and product design group is trying to figure out how to make your next Happy Meal toy, laptop, or Ikea table look like it was hand-made by a *MAKE* reader, recycled from scrap, sold on Etsy...? Will we soon have Potemkin crafters whose fake, procedurally-generated pictures, mottoes, and logos grace each item arriving from an anonymous overseas factory? Will the 21st-century equivalent of an offshore call-center worker who insists he is "Bob from Des Moines" be the Guangzhou assembly-line worker who carefully "hand wraps" a cell-phone sleeve and inserts a homespun anticorporate manifesto (produced by Markov chains fed on angry blog-posts from online maker forums) into the envelope?

I wouldn't be surprised. Our species' capacity to commodify everything—even the anticommodification movement—has yet to meet its match. I'm sure we'll adapt, though: we could start a magazine for hobbyists who want to set up nostalgic mass-production assembly lines that use old-fashioned injection moulders to stamp out stubbournly identical objects in reaction to the corporate machine's insistence on individualized, 3D-printed, fake artisanship.

Close Enough for Rock 'n' Roll

I once gave a (now-notorious) talk at Microsoft Research about Digital Rights Management (http://craphound.com/msftdrm.txt) where I said, in part, "New media don't succeed because they're like the old media, only better: they succeed because they're worse than the old media at the stuff the old media is good at, and better at the stuff the old media are bad at."

I'd like to take that subject up with you today. Specifically, I'd like to examine it in light of the ancient principle of "Close enough for rock 'n' roll," and all that that entails.

What, exactly, does "close enough for rock 'n' roll" *mean?* Does it mean that rock 'n' roll isn't very good, so it doesn't matter if the details are a little fuzzy? I say no. I say that "close enough for rock 'n' roll" means: "Rock 'n' roll's virtue is in its exuberance and its accessibility to would-be performers. If you want to play rock 'n' roll, you don't need to gather up a full orchestra and teach them all to read sheet music, drill them with a conductor, and set them loose in a vaulted hall. Instead, you can gather two or three friends, teach them to play a I-IV-V progression in 4/4 time, and make some fantastic *noise.*"

Rock 'n' roll has two important virtues relative to orchestral music:

1. It costs a lot less to make, and so it costs less to make experimental mistakes

2. More people can participate in it, and can bring more experimental ideas to the field (see 1)

On the other hand, it lacks a lot of the important virtues of orchestral music: the sheer majesty of all that tightly coordinated virtuosity, the subtleties and possibilities opened up by having so many instruments in one place and available to be combined in so many ways.

In other words, rock 'n' roll is cheap, experimental, and fluid, and devotes most of its energy into the production of music. Orchestral music is expensive, formal, and majestic, but tithes a large portion of its effort to coordination and overheads and maintenance.

If the internet has a motif, it is rock 'n' roll's Protestant Reformation thrashing against the orchestral One Church. Rock 'n' roll gets lots of wee kirks built in every hill and dale in which parishioners can find religion in their own ways; choral music erects majestic cathedrals that humble and amaze, but take three generations of laborers to build.

The interesting bit isn't what it costs to *replicate* some big, pre-internet business or project.

The interesting bit is what it costs to do something *half as well* as some big, pre-internet business or project.

Take *Newsweek*. If you wanted to launch *Newsweek* today, you'd probably have to spend as much as *Newsweek* did. Maybe more, since you'd not only have to do what *Newsweek* does, you'd have to somehow outspend or outmaneuver *Newsweek* to get there.

But what does it cost to publish something half as good as *Newsweek*, say, the *Huffington Post*? Sure,

HuffPo has brought in about $20MM in venture capital, but ignore that sum—that's how much they can sweet talk out of the world of finance. I'm talking about how much capital it cost to build and operate *HuffPo*. A tiny, unmeasurable fraction of what it cost to build and run *Newsweek*.

But *HuffPo* is at least half as good as *Newsweek*—in audience reach, in influence, in news quality, in return-on-investment (though not in absolute profitability—that is, a dollar put into *HuffPo* will generate more income than a dollar put into *Newsweek*, but *HuffPo* uses a lot fewer dollars than *Newsweek* does, and returns fewer dollars in total than *Newsweek*, too).

What's more, as time goes by, we can expect it to get *cheaper* to get more *Newsweek*-like. Cheaper and better ad-sales markets. Larger pools of interested people with the time and skill and tools to follow breaking news. Even cheaper printing and logistics, should *HuffPo* go hardcover, thanks to the spread of cheap printer-binders around the world.

This is the pattern: doing something x percent as well with less-than-x percent of the resources. A blog may be 10 percent as good at covering the local news as the old, local paper was, but it costs less than 1 percent of what that old local paper cost to put out. A home recording studio and self-promotion may get your album into 30 percent as many hands, but it does so at 5 percent of what it costs a record label to put out the same recording.

What does this mean? Cheaper experimentation, cheaper failure, broader participation. Which means more diversity, more discovery, more good stuff that could never surface when the startup costs were so high that no one wanted to take any risks.

What's driving this cost-reduction? Part of it is the free ride on general technological development. Everyone—even the big, lumbering, expensive companies—needs cheaper hard drives, cheaper networks, cheaper computers. Every society is trying to increase the general technical literacy of its population, because every employer benefits from technical literacy in its workforce.

Partly, it's a free ride on overinvestment bubbles. When the dotcoms came along, they were—canonically—founded by two hackers in a garage working on doors balanced on sawhorses. They were so humble in origin that it was easy to believe that they'd grow to three or four hundred times their present size. Even three or four *thousand* times their present size. So they attracted capital—who doesn't like a crack at a 4,000X payout? More capital than they could absorb—because buying more sawhorses and doors and garages and commodity servers just doesn't cost that much. With all that money came a burden to spend, to try to grow a business large enough to pay off all that investment, which meant luring great numbers of bright people into the startup world, training them as you went on technical matters, turning them into internet people.

When the overinvestment bubbles (dotcom, finance) crashed, you were left with a lot of skilled smart people, a lot of equipment that had gotten cheap fast thanks to enormous consumption by overfinanced companies. This, too, made it cheaper to start something new.

But even without overinvestment, the gap between rock 'n' roll and the orchestra is narrowing. Technology is giving us the organizational equivalent of a really kick-ass synthesizer, one that can allow a one-man band to sound like a whole firm. It may be that we'll never get to a point where you could build Disneyland today for one tenth of what Disney has spent since 1955. But I'm pretty sure that in my lifetime, you'll be able to build an 80 percent Disneyland (you could call it "Disneyla") for maybe 30 percent of the capital sunk into the Magic Kingdom.

This is one of the great conundra of our era: the spectre that haunts every executive, every government, every powerful person who owes her stature to her command of an empire that enjoys its pride of place thanks to the prohibitive cost of replicating it.

But lurking in those 80 percent replacements are an infinitude of ideas too weird and too funky and implausible to try at full price. Lurking there are ideas as weird and dumb as a company called (I kid you not) Google, an encyclopedia that everyone can write, a wireless network standard based on open spectrum that anyone is allowed to use, without central planning.

It's rock 'n' roll, and if it's too loud, you're too old.

About the Author

Cory Doctorow (www.craphound.com) is a science-fiction novelist, blogger, and technology activist. He is the co-editor of the popular weblog Boing Boing (www.boingboing.net) and a contributor to the *Guardian*, the *New York Times*, *Publishers Weekly*, *Wired*, and many other newspapers, magazines, and websites. He was formerly Director of European Affairs for the Electronic Frontier Foundation (www.eff.org), a nonprofit civil liberties group that defends freedom in technology law, policy, standards, and treaties.

His novels are published by Tor Books and HarperCollins UK and simultaneously released on the Internet under Creative Commons licenses that encourage their reuse and sharing, a move that increases his sales by enlisting his readers to help promote his work. He has won the Locus and Sunburst awards, and has been nominated for the Hugo, Nebula, and British Science Fiction awards.

Doctorow's young adult novels include *New York Times* bestseller *Little Brother*, and its follow-up *For the Win*. Tachyon published the first collection of his essays, *Content: Selected Essays on Technology, Creativity, Copyright, and the Future of the Future*; IDW published a collection of comic books inspired by his short fiction, *Cory Doctorow's Futuristic Tales of the Here and Now*. His latest adult novel is *Makers*, and his latest short-story collection is *With a Little Help*. His forthcoming books include *Pirate Cinema*, *Rapture of the Nerds*, and *The Great Big Beautiful Tomorrow*.

On February 3, 2008, Cory Doctorow became a father. The little girl is called Poesy Emmeline Fibonacci Nautilus Taylor Doctorow and is a marvel that puts all the works of technology and artifice to shame.